WORKBOOK

T0268944

Cambridge IGCSE™

Biology
Practical Skills

Salma Siddiqui

HODDER
EDUCATION
AN HACHETTE UK COMPANY

We have carried out a health and safety check of this text and have attempted to identify all recognised hazards and suggest appropriate cautions. However, the Publishers and the authors accept no legal responsibility on any issue arising from this check; whilst every effort has been made to carefully check the instructions for practical work described in this book, it is still the duty and legal obligation of schools to carry out their own risk assessments for each practical in accordance with local health and safety requirements.

For further health and safety information (e.g. Hazcards) please refer to CLEAPSS at www.cleapss.org.uk

The Publishers would like to thank the following for permission to reproduce copyright material.

Photo credits

p.10 © Sciencephotos/Alamy Stock photo **p.14** / © Dr Keith Wheeler/Science Photo Library; **p.112** © UCLES; **p.124** © Biophoto Associates/Science Photo Library.

Acknowledgements

Cambridge International copyright material in this publication is reproduced under licence and remains the intellectual property of Cambridge Assessment International Education.

Cambridge Assessment International Education bears no responsibility for the example answers to questions taken from its past question papers which are contained in this publication.

Third-party websites and resources referred to in this publication have not been endorsed by Cambridge Assessment International Education.

Every effort has been made to trace all copyright holders, but if any have been inadvertently overlooked, the Publishers will be pleased to make the necessary arrangements at the first opportunity.

Although every effort has been made to ensure that website addresses are correct at time of going to press, Hodder Education cannot be held responsible for the content of any website mentioned in this book. It is sometimes possible to find a relocated web page by typing in the address of the home page for a website in the URL window of your browser.

Hachette UK's policy is to use papers that are natural, renewable and recyclable products and made from wood grown in well-managed forests and other controlled sources. The logging and manufacturing processes are expected to conform to the environmental regulations of the country of origin.

Orders: please contact Hachette UK Distribution, Hely Hutchinson Centre, Milton Road, Didcot, Oxfordshire, OX11 7HH. Telephone: +44 (0)1235 827827. Email education@hachette.co.uk Lines are open from 9 a.m. to 5 p.m., Monday to Friday. You can also order through our website: www.hoddereducation.com

ISBN: 978 1 3983 1046 9

© Salma Siddiqui 2021

First published in 2021 by
Hodder Education,
An Hachette UK Company
Carmelite House
50 Victoria Embankment
London EC4Y 0DZ

www.hoddereducation.com

Impression number   10 9 8 7 6 5 4 3 2 1

Year   2024 2023 2022 2021

All rights reserved. Apart from any use permitted under UK copyright law, no part of this publication may be reproduced or transmitted in any form or by any means, electronic or mechanical, including photocopying and recording, or held within any information storage and retrieval system, without permission in writing from the publisher or under licence from the Copyright Licensing Agency Limited. Further details of such licences (for reprographic reproduction) may be obtained from the Copyright Licensing Agency Limited, www.cla.co.uk

Cover photo © Werner Dreblow – stock.adobe.com

Illustrations by Aptara, Inc.

Typeset in Minion Pro 11/14 pts. by Aptara, Inc.

Printed and bound by CPI Group (UK) Ltd, Croydon, CR0 4YY

A catalogue record for this title is available from the British Library.

Contents

Note that there are no practicals provided in this book for the following topics: Human nutrition, Diseases and immunity, Excretion in humans, Drugs, Inheritance, Organisms and their environment and Human influences on ecosystems, as there are no traditional practicals in these areas. Due to this omission, we have therefore renumbered the sections in this book, so they may differ from the numbered sections in the accompanying Student's Book and Workbook.

Introduction

How to use this book

This *Practical Skills Workbook* will help you keep a record of the practicals you have completed, as well as your results and conclusions. It covers the Cambridge IGCSE™ and IGCSE (9–1) Biology syllabuses (0610/0970) for examination from 2023. This resource provides additional practice for the practical skills required by the syllabus with a focus on the investigation-focused learning objectives. These investigative learning objectives are covered more fully in the accompanying Student's Book in this series.

This resource covers Core and Supplement content. Supplement investigations and questions are indicated by a lined box around the text, as shown below.

> **4** Convert the typical diameter of an animal cell to µm.
>
> ....

Some practicals also have Going Further sections, which provide additional questions that apply the scientific theory learned from the practical to different contexts. These go beyond Core and Supplement level and can be used as stretch activities.

GOING FURTHER

Why do you think it is important to know the food type of different foods?

....

....

Completing the investigations

At the start of each investigation we have provided a brief piece of context to help explain how the science behind the practical ties into the wider syllabus. Key terms and equations that are relevant to each investigation are also provided.

The aim of each practical is then laid out, along with a list of apparatus needed to complete the practical as suggested. Your teacher will inform you if they have decided to change any of the equipment and if the method needs to be adapted as a result.

Before you begin the practical and start on the method it is vital that you read and understand the safety guidance, and take any necessary precautions. Once you have carried out a risk assessment and made everything safe, you should check with your teacher that it is appropriate to begin working through the method.

The method itself is presented in a step-by-step fashion and you should read it through at least once before starting, making sure you understand everything. Then, ensuring that you don't miss anything out, you should work through the practical safely. Tips may be provided to help with particularly problematic steps.

Questions and answers

Within each practical there are clear sections laid out for observations where you should record your results as you complete the practicals. Scaffolded questions are also provided to help you develop conclusions and evaluate the success of the experiment.

At the end of the book are past paper questions, which relate to the practicals within this book and provide useful practice. Your teachers may decide to set this as part of the lesson, or as practice at a later date.

Sample answers to all of the questions are provided in the accompanying *Cambridge IGCSE™ Biology Teacher's Guide with Boost Subscription*.

Experimental skills and abilities

Skills for scientific enquiry

The aim of this book is to help you develop the skills and abilities needed to perform practical laboratory work accurately and safely. We start by highlighting the importance of working safely. We then look at how to plan an experiment and introduce the apparatus and measuring techniques that you will use most often. Next we show you how to take and record measurements accurately, and how to make drawings of biological specimens. Finally, we will discuss how to carry out and evaluate an investigation. You should then be ready to work successfully through the experiments and laboratory activities that follow.

Special note to teachers

We believe that the suggested experiments can be carried out safely in school laboratories. However, it is the responsibility of the teacher to make the final decision depending on the circumstances at the time. Teachers must ensure that they follow the safety guidelines set down by their employers, and a risk assessment must be completed for *any* experiment that is carried out. Teachers should draw students' attention to the hazards involved in the particular exercise to be performed. The hazards are shown within the 'Safety guidance' section of each practical.

Safety

The science laboratory is potentially a dangerous place. Although any potential hazards are pointed out in the 'Safety guidance' at the start of each investigation, remember, it is your responsibility to be vigilant about any hazards, not only to yourself but to others in the laboratory. You should treat all chemicals as hazardous and handle them with care. Some are also flammable, such as alcohol (ethanol). Follow your teacher's instructions and warnings.

Here are some precautions to help ensure your safety when carrying out experiments in the laboratory.

1 **Always wear shoes** to protect your feet if a heavy weight should fall on them.

2 **Tie back long hair** to prevent it being caught in a flame.

3 **Personal belongings** – leave them in a sensible place so that no one will trip over them.

4 Wear a lab coat at all times.

5 **Protect eyes and skin from contact with any chemicals.** Wear eye protection and/or gloves when working with chemicals and always follow your teacher's instructions.

6 Keep your hands away from your face, eyes and mouth when working with chemicals and biological specimens. This includes not applying cosmetics, not adjusting contact lenses and not biting your fingernails. **Never eat or drink in the laboratory.**

7 If any chemicals splash into your eyes, immediately go to the nearest sink and flush your eyes with water. Report to staff.

SAFETY GUIDANCE

Familiarize yourself with the following hazard key:

C = Corrosive substance

H = Harmful or irritant

T = Toxic

F = Flammable

B = Biohazard

O = Oxidizing

N = Harmful to environment

8 Dispose of chemicals and biological specimens safely, as instructed.

9 Report all accidents, spills, breakages or injuries to the teacher.

10 **Hot liquids and solids** – handle with caution to avoid burns; set in a safe position where they will not be accidentally knocked over.

11 **Bunsen flames and flammable liquids** – turn the Bunsen burner off when not in use. Make sure the Bunsen flame is out before handling flammable liquids, such as alcohol (ethanol).

12 Do not allow electrical equipment to come into contact with water.

13 Be particularly careful with glassware, which can break easily, and with scalpels and sharp knives, which can cause nasty cuts.

14 Always wash your hands before leaving the laboratory.

Planning investigations

When preparing a plan to answer a specific question or extend a method to a new situation, you should produce a logical and safe procedure.

- You should identify the variables in the investigation and decide which ones to investigate and which ones you should try to keep constant. The variable that is changed is known as the **independent variable**. The variable that is measured is known as the **dependent variable**. To discover the relationship between variables, you should change only one variable at a time.
- Once you know what you will need to measure, you can decide on the apparatus and materials to be used. You should ensure that you choose measuring devices that have sufficient precision.
- Before you write a plan, familiarise yourself with how to use the apparatus and develop a plan of work. It will be helpful to decide how to record your results; draw up tables in which to record your measurements, if appropriate.
- Describe how you would carry out the experiment and include any safety precautions. It is useful to include a sketch of the experimental set-up.
- You may be asked to suggest how you would improve your experiment or make it more reliable. Taking more readings and repeating the experiment to see if you obtain consistent results is a good suggestion.

Controls

Many investigations require a control to prove that any reactions occurring in the investigation are, in fact, due to the stated cause and not due to other factors. For example, if you wished to see if what you observed was due to the presence of the enzyme you could substitute the enzyme in your control test tube with water or buffer. All other variables, such as substrate concentration, temperature, volume, pH, timing, should remain constant. The control will then provide a reference point of comparison for the main investigation.

pH indicators

pH of a solution tells you if a solution is acidic or alkaline. The pH scale ranges from 0–14, with pH 7 being neutral. An indicator is a substance that changes colour when it is added to acidic or alkaline solutions. Litmus paper is used to test if a solution is acidic or alkaline. It comes as red litmus paper or blue litmus paper. Red litmus paper remains red in acidic solutions and turns blue in alkaline solutions. It turns purple in neutral solutions. Blue litmus paper turns red in acidic solutions but remains blue in neutral or alkaline solutions.

Universal indicator solution or paper can show how strongly acidic or alkaline a solution is. Dipping the universal indicator paper strip in the solution changes its colour according to the pH of the solution. The pH can be read by matching the colour with the colour on the packet.

Using and organising techniques, apparatus and materials

You will be provided with appropriate equipment, such as balances, syringes, pipettes, measuring cylinders and rulers. You should choose the most suitable measuring device for the task.

Measuring accurately and precisely

It is important to learn to use equipment to measure consistently and accurately to obtain the best possible results. Accuracy and precision will depend on your ability to use the equipment properly and also depend on the equipment used to make the measurement – the smallest division of the scale. For example, if you needed to measure a volume of $10\,cm^3$ you will get a more accurate result using a $10\,cm^3$ measuring cylinder than a $100\,cm^3$ measuring cylinder or a beaker. This is because any volume measured with the $10\,cm^3$ measuring cylinder will be measured to the nearest small scale division, which is $0.1\,cm^3$, but the smallest scale division on a $100\,cm^3$ measuring cylinder is $1\,cm^3$, and on a beaker it is likely to be 10 or $20\,cm^3$.

Measuring volumes

Measuring cylinders, beakers and syringes are used to measure volumes of liquids. The SI units of cm^3 and dm^3 should be used for measuring liquids. However, some old glassware is marked in millilitres (ml), where:
- 1 millilitre = $1\,cm^3$
- 1 litre = $1000\,cm^3 = 1\,dm^3$.

When making a reading, the measuring cylinder should be vertical and your eye should be level with the bottom of the curved liquid surface – the meniscus (see Figure 1). Stand the measuring cylinder on a level surface and take the reading with your eye level with the bottom of the meniscus. (For mercury, the meniscus formed is curved oppositely to that of other liquids and you should read the level at the top of the meniscus in a mercury thermometer.)

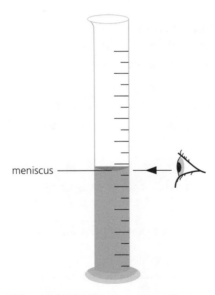

Figure 1 Reading a measuring cylinder

Measuring mass

A **balance** is used to measure the mass of an object. There are several types available. The one that you are most likely to use in your lab is a digital balance (see Figure 2), which gives a direct reading of the mass placed on the pan.

- The unit of mass is the kilogram (kg).
- The gram (g) is one-thousandth of a kilogram: $1\,g = \dfrac{1}{1000}\,kg = 10^{-3}\,kg = 0.001\,kg$.
- The smallest mass that can be measured on the scale setting you are using is probably 1 g or 0.1 g. This is the precision of the balance.

Figure 2 A digital top-pan balance

Measuring length

A **ruler** is often used to measure length. The unit of length is the metre (m). Submultiples are:

- 1 centimetre (cm) = $10^{-2}\,m$
- 1 millimetre (mm) = $10^{-3}\,m$
- 1 micrometre (μm) = $10^{-6}\,m$
- 1 kilometre (km) = $10^{3}\,m$

The correct way to measure with a ruler is shown in Figure 3, with the ruler placed as close to the object as possible. Your eye must be directly above the mark on the scale or else the thickness of the ruler causes parallax errors. The precision of the ruler (the smallest distance on the scale) is 1 mm.

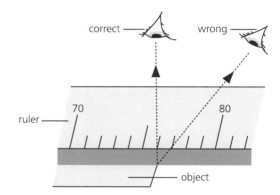

Measuring time intervals

Clocks, watches and timers can be used to measure time. The unit of time is the second (s).

To avoid mistakes in calculations, record times in the smallest units; for example, seconds rather than in minutes and seconds. Convert to the appropriate units if the answer requires it.

Figure 3 This reading is 76 mm or 7.6 cm

When using a stopwatch, reaction times may influence the reading. Human reaction time varies but is likely to be about 0.5 s. The reading on the stopwatch may be to the nearest 0.1 s or even 0.01 s, but the timing is only accurate to about 0.5 s.

For very short time intervals, a more accurate result may be obtained by measuring longer time intervals; for example, time your pulse rate over 60 seconds rather than over 15 seconds. A measurement error of 0.5 s has a smaller effect on a longer time interval.

1 What apparatus should you use to measure each of the following?

    a 100 cm³ water ....

    b 1 cm³ glucose solution ....

    c width of a plant stem ....

    d time taken to run across the lab ....

    e 5 g sugar ....

Changing measurements

- Take readings more frequently if values are changing rapidly.
- It will often be helpful to work with a partner who watches the timer and calls out when to take a reading.
- Pressing the lap-timer facility on the stopwatch at the moment you take a reading freezes the time display for a few seconds and will enable you to record a more accurate time measurement.
- For rapidly changing measurements, it may be necessary to use a **datalogger** and computer.

Observing, measuring and recording

Make a list of the apparatus you will use in an experiment and record the smallest division of the scale of each measuring device.

Having collected together and familiarised yourself with the equipment and materials needed for an experiment, you are now ready to start making some observations and measurements.

- You should record any difficulties encountered in carrying out the experiment with suggestions on how to overcome them. Any precautions taken to achieve accuracy in your measurements should also be recorded.
- Do not dismantle the equipment until you have completed the analysis of your results and are sure you will not have to repeat any measurements!

Consider:

- how many significant figures your data will have
- how you will record your results.

Significant figures

- The number of digits given for a value in a measurement or calculated value indicates how accurate we think it is. This is called the number of **significant figures**. A measurement written as $5\,cm^3$ has one significant figure but a measurement written as $5.0\,cm^3$ has two significant figures. The 0 after the decimal point is significant because it shows we have measured a volume to the nearest $0.1\,cm^3$.
- When doing calculations your answer should have the same number of significant figures as the measurements used in the calculation. For example, if your calculator gives an answer of 1.23578, this would be 1.2 if your measurements have two significant figures and 1.24 if your measurements have three significant figures.
- To round a number to two significant figures, look at the third digit. Round up if the digit is 5 or more, and round down if the digit is 4 or less. So 1.23 rounds to 1.2 but 1.25 rounds to 1.3.
- If a number is expressed in standard notation, the number of significant figures is the number of digits before the power of 10. For example, 6.24×10^2 has three significant figures.
- If values with different numbers of significant figures are used to calculate a quantity, quote your answer to the smallest number of significant figures.

Sources of error

Every measurement of a quantity is an attempt to find its true value and is subject to errors arising from the limitations of the apparatus and the experimental procedure.

Systematic errors

An error that makes your measurements consistently too high or too low is called a **systematic error**. For example:

- When reading the volume of a measuring cylinder, you should ensure that your eye is directly opposite the bottom of the meniscus or else your measurements of volume will consistently be too low or too high.
- When using a rule to measure a height, the rule must be held so that it is vertical. If it is at an angle to the vertical then a systematic error will be introduced.
- If a top-pan balance is not set to 0 ('tared') before making a measurement, this produces an error in the mass you are measuring. For example, if you are weighing a chemical on a piece of filter paper, place the filter paper on the balance and tare the balance so that it reads 0 with the filter paper on the balance, then add the chemical to be weighed on the filter paper. If you are not able to tare the balance, record the weight of the filter paper, add the chemical, and subtract the weight of the filter paper from the final weight.
- Check for any zero error when using a measuring device. If it cannot be eliminated, correct your readings by adding or subtracting the zero error to them.

Random errors

Random errors usually result from not taking the same measurement in exactly the same way, for example, judging when a colour change has occurred. This type of error produces readings that vary randomly above or below the true value. Being consistent in your procedure, taking repeated measurements and then finding a mean value reduces the effect of random errors.

Tables

If several measurements of a quantity are being made, draw up a **table** in which to record your results.

- Think about the number of columns and rows you will need. Make sure to include a column for a mean value if you plan to take several repeated measurements.
- Use the column headings, or the first cell of the rows, to name the measurement and state its unit.
- Never add units in the body of the table.
- Repeat the measurement of each observation, if possible, and record the values in your table. If repeat measurements for the same quantity are significantly different, take a third reading. Calculate the mean (average) value from your readings.
- Numerical values should be given to the number of significant figures appropriate to the measuring device.

Table 1 Using a table to record measurements

Temperature/°C	Time taken for colour change/s		
	Repeat 1	Repeat 2	Mean
20	168	192	179
30	135	127	131
40	83	89	86

Recording observations by drawing

You are often required to record your observations of biological specimens or observations of plant and animal cells by drawing. Here are some tips to help you:

1 Keep a good selection of sharp pencils, a good eraser, a ruler and a sharpener.

2 Pretend that you are drawing for someone who has never seen this specimen. They should know exactly what it looks like, and how large it is, without seeing the specimen.

3 Start your drawing in the centre of the page. Use most of the available space but leave space on either side for labels.

4 Using a soft, sharp pencil, draw faint, clear, smooth lines. Avoid hesitant or broken lines. If you do draw in such a manner, make sure to rub these out and go over them in smooth, continuous lines.

5 Observation and attention to detail is the key to a good drawing. Add the main structures in the correct locations and proportions.

6 Make sure that junctions between lines are properly drawn; for example, if you are drawing a plant, observe exactly how the leaves are attached to the main stem. Is there a little stalk joining the main leaf blade to the stem or does the blade join directly on to the stem? Look at the exact arrangements of the leaves. Are the two leaves arranged exactly opposite each other or are they alternating?

7 Draw exactly what you see and not what you expect to see, and do not copy from a textbook.

8 Do not use shading or coloured pencils.

9 Once you are satisfied with your effort, go over it with firmer, darker lines, rubbing out any mistakes or fuzzy, broken lines. Keep your lines thin.

10 Label your drawing to show the main structures that are visible.

Labelling

1 Do not write the labels too close to the drawing.

2 Write labels clearly using a sharp pencil on either side of the drawing. Draw label lines with a ruler pointing precisely to the structure it labels. Do not put arrows on the lines.

3 Label lines should never cross each other.

4 You might be asked to annotate a label, for example, to give a brief description of structures or functions.

Plan drawing

A plan drawing is a special form of recording that shows only the outlines of tissues, and no individual cells (Figure 4).

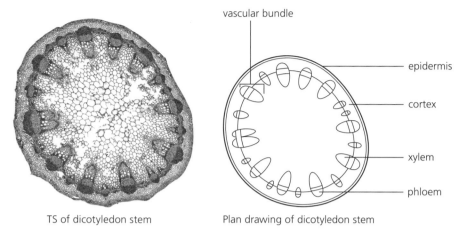

TS of dicotyledon stem   Plan drawing of dicotyledon stem

Figure 4 A dicotyledon stem

Magnification of drawing

You may choose to draw the same image at a size of 10 mm or 10 cm. Therefore, it is essential to work out and record the magnification on all your drawings.

To calculate the magnification, convert all values to the same unit, usually millimetres or micrometres. For example, if the actual length of a specimen is 12 mm but the length in your drawing is 36 mm:

$$\text{magnification} = \frac{\text{size of drawing (mm)}}{\text{actual size of object (mm)}}$$

$$= \frac{36}{12} = 3$$

magnification = ×3

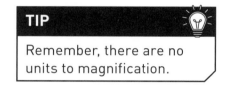

TIP

Remember, there are no units to magnification.

Handling experimental observations and data

Now that you have collected your measurements, you will need to process them. Perhaps there are calculations to be made or you will decide to draw a graph of your results. Then you can summarise what you have learnt from the experiment, discuss sources of experimental error and draw some conclusions from the investigation.

Consider:

- What is the best way to process your results?
- Are there some inconsistent measurements to be dealt with? **Anomalous data** are readings that do not fit the pattern of the other measurements.
- What sources of experimental error are there?
- What conclusions, generalisations or patterns can you draw?

Calculations

You may have to produce a mean (average) value to process your results. When calculating the mean of repeated measurements, ignore any anomalous values.

The **mean** is found by adding the values for a quantity you have measured, and dividing the sum by the number of values.

For example, if you measured the length of a branch as 81.5 cm and 81.6 cm, then:

$$\text{mean} = \frac{(81.5 + 81.6)}{2} = \frac{163.2}{2}$$

$$= 81.55 \text{ cm or } 81.6 \text{ cm}$$

The value has been given to three significant figures because that was the accuracy of the individual measurements.

The **range** of a set of measurements is the difference between the maximum and minimum values. For example, in Table 1 earlier, the range of temperatures investigated is 20 to 40 °C. The range in the measured times is 83 to 192 s. You can also describe the range of the repeat measurements. The range in the results at 20 °C is from 168 to 192 s.

Graphs

Graphs can be useful in finding the relationship between two quantities.

- You will need at least six data points, taken over as large a range as possible, to plot a graph, in order to see a trend.
- Choose scales that make it easy to plot the points and use as much of the graph paper as possible.
- Make sure you label each axis of the graph with the name and unit of the quantity being plotted (Figure 5).
- Mark the data points clearly with a dot within a circle (⊙) or a cross (×), using a sharp pencil.
- Join data points with a ruled line.
- The line of best fit is then drawn through them, if appropriate (as in Figure 5). In practice, points plotted on a graph from actual measurements may not lie exactly on a straight line or curve due to experimental errors.
- If possible, repeat any anomalous measurements to check that they have been recorded properly, or try to identify the reason for the anomaly.

Table 2 shows one student's experimental data from an investigation of the effect of light intensity on the rate of photosynthesis. Light intensity was varied by moving a lamp closer to a plant. The gas that was given off as a result of photosynthesis during 2 minutes at each distance was collected and measured.

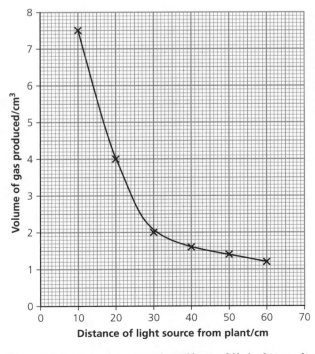

Table 2 Data values of the effect of light intensity on the rate of photosynthesis

Distance of light from the plant/cm	Volume of gas produced/cm³
10	7.5
20	4.0
30	2.0
40	1.8
50	1.5
60	1.3

Figure 5 Graph showing the effect of light intensity on the rate of photosynthesis

Conclusions

Once you have analysed your experimental results, summarise your conclusions clearly and relate them to the aim of the experiment.

- State whether a hypothesis has been verified. If your results do not, or only partially, support a hypothesis, suggest reasons why.
- If a numerical value has been obtained, state it to the correct number of significant figures. Compare your results with known values, if available, and suggest reasons for any differences.
- State any relationships discovered or confirmed between the variables you have investigated.
- Mention any patterns or trends in the data.

Evaluating investigations

Finally, evaluate your experiment and discuss how it could be improved. Could some things have been done better? If so, suggest changes or modifications that could be made to the procedure or the equipment used in the investigation. For example:

- Should repeat measurements be made?
- Are there enough results to show a pattern?
- Was the range of the independent variable good enough?
- Should you get further data in-between values; for example, if there is an uncertainty about the results in one part of the range?

Identify and comment on sources of error in the experiment. For example, it may be very difficult to eliminate all energy losses to the environment in an experiment where the temperature change of a liquid is measured; if that is the case, say so. Mention any sources of systematic error in the experiment, or random errors, and what might have caused them.

Characteristics and classification of living organisms

1.1 Dichotomous keys

A key allows you to use the general features of organisms to identify a specimen or to place it in the correct group. A dichotomous key presents two contrasting choices at each step. Each step eliminates some possible answers until you can correctly identify your specimen.

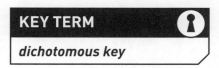

KEY TERM

dichotomous key

Aim

To produce a dichotomous key for the identification of tree leaves.

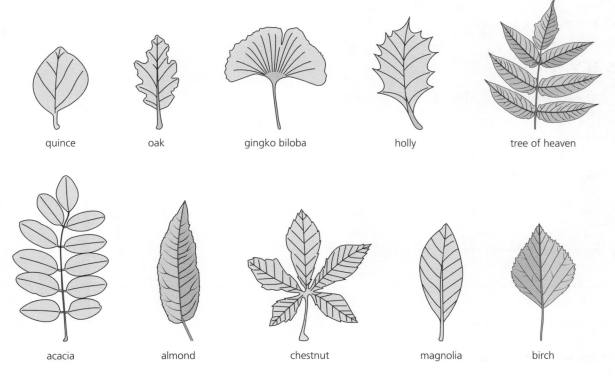

| quince | oak | gingko biloba | holly | tree of heaven |

| acacia | almond | chestnut | magnolia | birch |

Figure 1 A range of leaves

Apparatus

- Paper
- Pencil
- Ruler
- The leaf drawings from Figure 1, or actual specimens of a variety of leaves

SAFETY GUIDANCE

- This practical presents minimal risk. However, standard laboratory safety rules still apply at all times.
- Ensure that you wash your hands after handling fresh leaves.

Method

1 Research examples of dichotomous keys using library resources or the internet.

2 Look at the similarities and differences of the leaves provided by your teacher, or those in Figure 1. Record these in the Observations section below.

3 Identify an obvious feature that divides the leaves in two approximately equal-sized groups. Use this feature to start your key. You can use less obvious, more detailed features to divide them into smaller groups later.

4 Produce a dichotomous key. You might find it helpful to produce a draft on a piece of rough paper first.

5 Draw your final key in the Observations section.

6 Test your key using the range of different leaves provided by your teacher, or using Figure 1.

TIP

Consider only one characteristic at a time and keep the two choices in each step simple, such as 'single leaf', not 'a single leaf'. Also, avoid questions that combine more than one characteristic.

TIP

You may find it easier to number the leaves 1–10, to avoid writing the names at each step.

Observations

1 List the similarities and differences of the leaves.

....

....

....

....

....

....

2 Present your completed dichotomous key below.

3 Use your key to identify each of the leaves provided by your teacher, or each of the leaves in Figure 1.

1 ....   6 ....

2 ....   7 ....

3 ....   8 ....

4 ....   9 ....

5 ....   10 ....

Conclusions

How effective is your key?

....

....

....

....

Evaluation

1 Does your dichotomous key identify all the leaves?

....

2 If not, try to think of how you can improve your key so that it identifies all the leaves.

....

....

....

3 List three rules that can help you to construct a dichotomous key.

....

....

....

GOING FURTHER

Why are dichotomous keys easier to use than keys with three branches (three possible answers) for each stage?

....

....

Organisation of the organism

2.1 Looking at plant cells

Plant cells are tiny, too small to be seen without a microscope. They are almost transparent and therefore difficult to see unless treated with dyes, called stains. Most plant cells contain cytoplasm, a cell wall, cell membrane, nucleus, vacuole and chloroplasts. In this practical, iodine solution is used to stain the cells, to make different structures inside the cells show up more clearly.

Aim

To use a light microscope to observe, draw and label a selection of plant cells and calculate the magnification of the drawing.

Apparatus

- Eye protection
- Fine forceps
- A small piece of onion
- Scalpel or knife
- Cutting board or tile
- Microscope slide and cover-slip
- Mounted needle
- Iodine solution in dropping bottle
- Light microscope
- Filter paper
- Transparent ruler

Method

1 Cut a 0.5 cm long piece from the onion.

2 Separate two of the thin layers of the onion.

3 Using forceps, peel off the inner translucent membrane between the layers. This is the epidermal layer (see Figure 1).

KEY TERM

magnification

KEY EQUATION

$$\text{magnification} = \frac{\text{image size}}{\text{actual size}}$$

SAFETY GUIDANCE

- Eye protection must be worn.
- Iodine solution causes stains and may be an irritant [H] to the eyes. Wear eye protection. Take care to avoid getting it on skin, clothes or equipment.
- Cover any cuts on your hands with waterproof dressings or wear gloves.
- Take care to avoid cutting yourself with the scalpel/knife. Always cut directionally away from yourself.

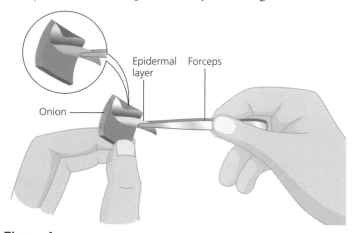

Epidermal layer   Forceps

Onion

Figure 1

4 Place two drops of iodine solution onto the centre of a microscope slide.

5 Place the epidermal tissue on top of the iodine solution, making sure it is flat and not folded. Avoid trapping air bubbles.

6 Carefully lower a cover-slip onto the slide. Do this by placing one edge of the cover-slip on the slide and using a mounted needle to lower the other edge down onto the slide.

7 Leave for five minutes while the iodine solution penetrates the cell membrane.

8 Soak up any liquid from around the edge of the cover-slip using some filter paper.

9 Place the slide on the microscope stage.

10 Select the lowest-power objective lens.

11 Looking through the eyepiece, turn the coarse-adjustment knob to increase the distance between the objective lens and the slide until the cells come into focus.

12 Rotate the fine-adjustment knob to bring the cells into a clear focus.

13 When you have found some cells, switch to a higher-power objective lens. Use only the fine focus dial to adjust the focus. Identify the nucleus, the cytoplasm surrounding the vacuole and the cellulose cell wall. Notice that the iodine solution stains the cytoplasm a light-yellow colour, but the nucleus takes up the stain more strongly than the cytoplasm.

14 Make a clear, labelled drawing of two or three epidermal cells and their component parts.

15 Measure and record the maximum length of two of the cells in your drawing.

16 Wash your hands thoroughly.

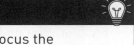

TIP

Always focus the microscope on the lowest-objective lens, first using the coarse adjustment, then the fine focus. Once you have done this, you only need to use the fine-adjustment knob to adjust the focus at high power.

TIP

When drawing cells look carefully at the shape of each cell and how the cells join with other cells. Draw a clear, continuous outline for each cell with no shading. The drawing should include a title that describes the specimen, and labels for each visible structure.

Observations

1 Draw and label a sample of two or three of the cells as seen under the microscope.

2 Record the length of two of the cells in your drawing.

Length of cell 1: .... mm

Length of cell 2: .... mm

Calculate the mean cell length in your drawing: .... mm

3 The typical length of onion epidermal cells is 0.25 mm.

Use this information to calculate the magnification of your drawing. Write the magnification underneath your drawing.

---

4 Convert the typical length of an onion epidermal cell into µm.

....

> **TIP**
>
> 1 mm = 1000 µm

Conclusions

The parts of plant cells that are clearly visible under the light microscope are

....

....

Evaluation

Complete the following sentences.

1 Care had to be taken when using iodine because

....

2 Iodine was used because

....

3 To view the plant cells clearly, the microscope objective needs to be

....

4 The cell membrane was present but was not visible because

....

5 Many plant cells contain green structures called chloroplasts. Onion epidermal cells do not have chloroplasts because

....

GOING FURTHER
• •

Research and name two other types of microscopes that help us to view extremely small cell structures that are too small to be seen using a light microscope.

....

....

2.2   Looking at animal cells

Almost all cells are microscopic. Animal cells are often smaller than plant cells. However, the nucleus, cytoplasm and cell membrane of animal cells can be seen using a light microscope. It is easy to remove some epidermal (skin) cells and look at them with a microscope.

Aim

To use a light microscope to observe, draw and label a selection of human epidermal (skin) cells and calculate the magnification of the drawing.

Apparatus

- Eye protection
- Transparent sticky tape
- Light microscope
- Microscope slide and cover-slip
- Soap
- 1% methylene blue stain in a dropping bottle
- Filter paper
- Mounted needle
- Forceps
- Transparent ruler
- Beaker of disinfectant (0.5–1% chlorine or bleach solution)

Method

1   Wash your wrist with soap and water.

2   Press a 2 cm piece of sticky tape onto the cleaned area of your skin.

3   Use the forceps to gently remove the tape. Be careful to avoid getting fingerprints on the tape.

4   Carefully place the tape, sticky-side up, on a clean microscope slide.

5   Add 2 drops of methylene blue solution to the tape.

6   Use the mounted needle to carefully lower a cover-slip over the sticky tape.

7   Soak up any excess stain from around the edge of the cover-slip using a piece of filter paper.

8   Place the slide on the microscope stage.

9   View the slide using the lowest-power objective lens. Turn the coarse-adjustment knob until the cells come into focus. Then use the fine-adjustment knob to bring the cells into a clear focus.

KEY TERMS

epidermis
stain

KEY EQUATION

$$\text{magnification} = \frac{\text{image size}}{\text{actual size}}$$

SAFETY GUIDANCE

- Eye protection must be worn.
- Wear a lab coat at all times. Wear gloves and handle all solutions with care. Avoid contact with skin, eyes or clothing. If solutions come into contact with your eyes or skin, wash immediately.
- Methylene blue causes stains and may be an irritant [H] to the skin and eyes.
- Dispose of your slides in the beaker of disinfectant provided. Take care to avoid spillage; it may be an irritant to skin and will stain your clothing.

10   Once you can see some cells, switch to a higher-power objective lens. Focus using the fine focus dial.

11   Find some intact cells and record your observations by making large labelled drawings of two or three cells.

12   Measure and record the width of the cells at their widest parts, in mm.

Observations

1   Draw and label a sample of two or three cells as seen under the microscope.

TIP

Do not write your text labels too close to the drawing. Instead, use a pencil and ruler to draw a line that links the text to the part that it labels, and which touches the structure. Do not put arrows on the lines. The label lines should not cross each other.

2   Record the width of two of the cells in your drawing.

Width of cell 1: ………….. mm

Width of cell 2: ………….. mm

Calculate the mean cell width in your drawing: …………. mm

3   The typical diameter of an animal cell is about 0.03 mm.

Use this to calculate the magnification of your drawing. Write the magnification underneath your drawing.

4   Convert the typical diameter of an animal cell to μm.

....

Conclusions

The parts of animal cells that are clearly visible under the light microscope are:

....

Evaluation

1 Compare these cells with the plant cells you drew previously in Experiment 2.1 'Looking at plant cells' on page 20.

Table 1

	Plant cells	Animal cells
Similarities		
Differences		

2 To view the cells clearly, the microscope objective needed to be

....

GOING FURTHER

Until recent updates to health and safety guidelines, students often used cells from the inside of their cheeks to observe animal cells. They took a sample by swabbing the inside of their cheek with a cotton swab.

1 Suggest the possible hazards of using cheek cells.

....

....

....

2 If you were allowed to use cheek cells, suggest how you could limit harm to other students.

....

....

3 Movement into and out of cells

3.1 Osmosis and water flow

A partially permeable cell membrane surrounds living cells. The cell membrane has tiny pores. Osmosis is the diffusion of water molecules from a high concentration of water molecules to a low concentration of water molecules through the partially permeable membrane. Visking dialysis tubing is an artificial partially permeable membrane. It is often used in practicals to demonstrate osmosis.

Aim

To demonstrate the effect of osmosis using Visking tubing.

Apparatus

- Eye protection
- Retort stand and clamp
- Capillary tube
- Visking dialysis tubing
- Elastic band
- 30% sugar solution with red dye
- Beaker with water
- Syringe
- Stopwatch
- Ruler
- Scissors

KEY TERMS

cell membrane
diffusion
osmosis
partially permeable

SAFETY GUIDANCE

- Eye protection must be worn.
- This practical presents minimal risk. However, standard laboratory safety rules still apply at all times.
- Treat glassware with care to avoid breakages and cuts.
- Take care with the red dye, it may stain skin and clothing.

Method

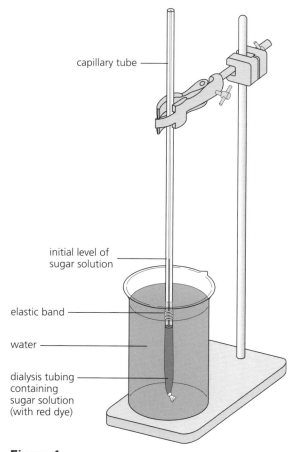

capillary tube

initial level of sugar solution

elastic band

water

dialysis tubing containing sugar solution (with red dye)

Figure 1

Cambridge IGCSE™ Biology Practical Skills Workbook

1 Cut a 15 cm length of dialysis tubing and soak in water for a few minutes.

2 Tie a knot tightly in one end of the dialysis tubing.

3 Use the syringe to fill the tubing with the red sugar solution.

4 Fit the tubing over the end of a capillary tube, ensuring that the sugar solution enters the capillary tube. Secure it in place with an elastic band.

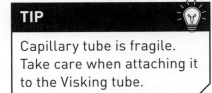

TIP

Capillary tube is fragile. Take care when attaching it to the Visking tube.

5 Clamp the capillary tube so that the dialysis tubing is totally covered by the liquid in the beaker, as in Figure 1.

6 Use a ruler to measure and record the starting level of the liquid in the capillary tube, and then start the stopwatch.

7 Record the level of liquid in the capillary tube after 10 minutes.

Observations

Level in capillary tube at start: ....

Level in capillary tube after 10 minutes: ....

Difference in levels: ....

Conclusions

Explain the change observed in the level of liquid in the capillary tube.

....

....

Evaluation

1 Explain why the red dye was added to the sugar solution.

....

2 Outline how this experiment could be improved to make the results more reliable.

....

....

....

3 Suggest what would happen if you replaced the capillary tube with
   a one with a smaller bore (hole).

....

   b one with a wider bore (hole).

....

Which method, (a) or (b), would give more accurate results?

....

....

4   Suggest what you think would happen if the liquid in the dialysis tubing was water and the liquid in the beaker was replaced with sugar solution. Give an explanation for your answer.

....

....

....

....

3.2   Observing osmosis in potato tissue

Plant cells have a cell wall outside the cell membrane. The cell wall allows diffusion of water and solute molecules into and out of the cell, but the cell membrane controls which molecules can move in and out of the cell. By measuring changes in the mass and length of a piece of potato before and after soaking it in water or in sugar solution, you can determine the direction of movement of water into or out of the potato.

Aim

To observe osmosis in plant tissue by observing its mass and length after soaking it in water or in sugar solution.

KEY EQUATION   ⊜
$$\text{percentage change} = \frac{\text{change in mass}}{\text{initial mass}} \times 100$$

Apparatus

- Eye protection
- Cork borer
- Scalpel or knife
- Large potato
- Cutting board or tile
- Ruler
- 2 test tubes
- 30% sugar solution
- Water
- Marker pen
- Balance
- Tissue paper
- Pencil

KEY TERMS   🛈
osmosis
percentage change

SAFETY GUIDANCE   ⚠
- Eye protection must be worn.
- Take care to avoid cutting yourself with the scalpel/knife. Always cut directionally away from yourself.
- Cork borers have sharp edges; take care not to cut yourself.

Method

1   Make a prediction of what will happen to the length and mass of potato cylinders when they are left to soak in water and in sugar solution. Record your prediction in the Observations section.

2   **Carefully** cut away the skin from the potato.

3 **Carefully** cut two potato cylinders of the same diameter using a cork borer, as shown in the diagram.

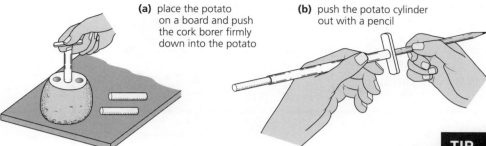

(a) place the potato on a board and push the cork borer firmly down into the potato

(b) push the potato cylinder out with a pencil

Figure 1

4 Cut the cylinders to the same length, ensuring they are at least 50 mm long.

5 Label two test tubes as 'water' and 'sugar'. Add the water to one test tube and the sugar solution to the other, so that they are half full.

6 Measure the length and mass of one potato cylinder and add to the tube labelled 'water'.

7 Draw a table to record your results in the Observations section.

8 Repeat step 6 with the other potato cylinder and add to the tube labelled 'sugar'. Record your measurements and leave the tubes overnight.

9 The next day, remove the cylinders from the test tubes and record your observations of the texture (firmness or flabbiness) of each potato cylinder.

10 Blot dry the cylinders with tissue paper and record their length and mass in your table.

11 Describe the changes in length and mass of the cylinders.

> **TIP**
>
> Use the smallest units (mm) to measure the cylinder lengths. If you mix measurements in centimetres and millimetres it is easy to make mistakes in calculations.

> **TIP**
>
> Your table should have enough columns and rows to record all data, with units for measurements in the headings.

Observations

1 Record your prediction before you begin.

....

....

2 Draw a table to record the length and mass of the potato cylinders in the two tubes.

3 Description of cylinders after soaking

   – in water: ....

   – in sugar solution: ....

   Changes in length and mass after soaking

   – in water: ....

   – in sugar solution: ....

4 Collect and record results for the changes in length and mass from at least two other groups in the class.

   ....

   ....

   ....

5 Suggest why the class results vary.

   ....

   ....

   ....

6 To make a more accurate comparison of the change in mass, the percentage change in mass can be calculated.

   Calculate the percentage change in mass for each cylinder.

Conclusions

1 Comment on your prediction.

   ....

   ....

2 How did the water and the sugar solution affect the length and mass of the potato cylinders? Use your understanding of osmosis to explain your findings.

....

....

....

....

....

....

....

....

Evaluation

1 Why was the skin removed from the potato in step 2?

....

....

2 Why was the cork borer used in step 3?

....

....

3 Why was it important to use cylinders of approximately the same length?

....

....

4 Describe how the method could be changed to improve the accuracy of this experiment.

....

....

5 The potato cylinders were blotted dry before recording their mass in step 10, but not before placing them in the test tubes (steps 6 and 8). Explain why.

....

....

....

6 Design an experiment to investigate the effects of different concentrations of sugar solution on potato cylinders. How could you use the results to estimate the concentration of sugar solution in the potato?

....

....

....

....

....

....

3.3 Osmosis and turgor

Plant cells have a cell wall outside the cell membrane. The cell wall is completely permeable and allows diffusion of water and solute molecules in and out of the cells, but the cell membrane is only partially permeable. If a partially permeable membrane separates two solutions, water will diffuse across the membrane from the dilute to the more concentrated solution by osmosis. Plant cells become turgid (swollen) when they take in a lot of water and the contents of the cell push out against the cell wall.

Aim

To model turgor in a plant cell using partially permeable Visking dialysis tubing.

Apparatus

- Eye protection
- Syringe
- Visking dialysis tubing
- Syrup or concentrated sugar solution
- Test tube
- Water
- Stopwatch
- Ruler
- Scissors

KEY TERMS

diffusion
osmosis
partially permeable
turgid/turgor
water potential

SAFETY GUIDANCE

- Eye protection must be worn.
- This practical presents minimal risk. However, standard laboratory safety rules still apply at all times.
- Treat glassware with care to avoid breakages and cuts.

Method

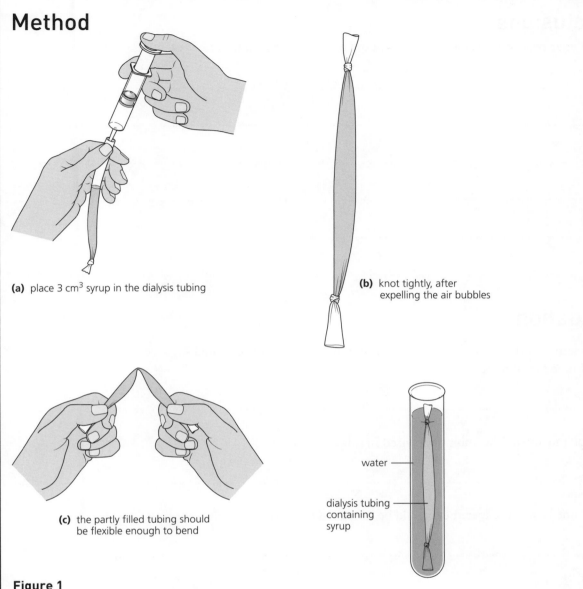

(a) place 3 cm³ syrup in the dialysis tubing

(b) knot tightly, after expelling the air bubbles

(c) the partly filled tubing should be flexible enough to bend

water

dialysis tubing containing syrup

Figure 1

1 Cut a 20 cm length of Visking dialysis tubing and soak it in water for a few minutes.

2 Tie a knot tightly at one end.

3 Using a syringe, place 3 cm³ of syrup or strong sugar solution in the tubing.

4 Tie a knot tightly in the open end of the tubing.

5 Record your observations on the floppiness of the Visking dialysis tubing.

6 Place the tubing in a test tube filled with water. Leave for 40 minutes.

7 After 40 minutes, remove the tubing and record your observations on the floppiness of the dialysis tubing.

Observations

Before leaving tubing in water: ....

After leaving tubing in water: ....

Conclusions

Explain your observations by using ideas about osmosis, turgor and water potential.

....

....

....

....

....

Evaluation

1 Explain why the dialysis tubing containing the syrup or concentrated sugar solution was left in the test tube for 40 minutes.

....

2 Explain why dialysis tubing was used in this experiment.

....

3 Outline how this experiment could be improved to make the results more reliable.

....

....

....

4 Predict, with a reason, how you think the floppiness/firmness of the dialysis tubing at the beginning and at the end of the experiment would compare if the sugar solution was more dilute. Describe how you could investigate this.

....

....

....

....

3.4 Plasmolysis

Plant cells have a cell wall outside the cell membrane. The cell wall is completely permeable and allows diffusion of water and solute molecules in and out of the cells, but the cell membrane is only partially permeable and only allows diffusion of water and some small molecules. When plant cells are placed in a solution that is more concentrated than in the cell, water diffuses of the cell out by osmosis and the cytoplasm shrinks away from the cell wall. This process is called plasmolysis and the cells become flaccid.

Aim

To observe plasmolysis in plant cells.

Apparatus

- Eye protection
- Small piece of red onion, cut from the inner layers
- Scalpel or knife
- Fine forceps
- Microscope
- Microscope slide and cover-slip
- Dropping pipette
- 30% sugar solution
- Blotting paper
- Mounted needle

Method

In this investigation you will be using a piece of red onion. The cells of the **outer epidermis** contain a red pigment called anthocyanin, so you will not need to add stain to see the contents of the cell.

1 Cut a piece about 0.5 cm long from the onion.

2 Using fine forceps, peel off the thin, red outer membrane (epidermis).

3 Place the onion epidermis on a microscope slide.

4 Add a drop of water and use the mounted needle to lower a cover-slip on top.

5 Focus on a small group of cells using the microscope and record your observations.

6 Draw a labelled diagram of what you observe. Not all the cells in red onion epidermis will contain pigmented cytoplasm.

7 Place a few drops of sugar solution on the slide, to one side of the cover-slip.

8 Use the blotting paper on the other side of the cover-slip to draw the solution across.

9 Observe under the microscope and record your observations.

10 Draw and label a diagram of what you see.

KEY TERMS

flaccid
osmosis
permeable
plasmolysis/plasmolysed
turgid/turgor
water potential

SAFETY GUIDANCE

- Eye protection must be worn.
- Take care to avoid cutting yourself with the scalpel/knife. Always cut directionally away from yourself.
- Care needs to be taken when using slides and the risks associated with broken glass.

Observations

1 Observations of cells mounted in water.

....

....

Draw and label the cells mounted in water.

2 Observations of cells mounted in sugar solution.

....

....

....

Draw and label the cells mounted in sugar solution.

Conclusions

1 How did placing the cells in sugar solution affect the cells?

....

....

2 Explain the effects of immersing plant cells in water and sugar solutions by using ideas about osmosis, turgor, plasmolysis and water potential.

....

....

....

....

....

3 Explain how you think a more concentrated sugar solution would affect the cells.

....

....

Evaluation

Suggest one way in which the strength of evidence from this experiment could be improved.

....

....

GOING FURTHER

1 Most land plants do not survive when watered with sea-water. Suggest why.

....

....

....

2 Design an experiment to test how much salt a plant can tolerate.

....

....

....

3 What is the independent variable in your investigation?

....

4 What variables would you need to control (keep the same)?

....

Biological molecules

4.1 Food tests

Foods contain substances such as carbohydrates (starch or sugars), proteins and fats. Different food types can be identified in the laboratory using a number of different food tests:

- Iodine solution test for starch
- Benedict's solution test for reducing sugar
- Biuret solution test for protein
- Ethanol emulsion test for fat and oil
- DCPIP test for vitamin C.

KEY TERMS

Benedict's solution
biuret test
control
DCPIP test
iodine solution
reducing sugar

Aim

To become familiar with the results of tests to confirm the presence of starch, reducing sugar, protein, fat and vitamin C.

Apparatus

- Eye protection
- Spatula
- Test tubes in a test-tube rack
- Water
- Tongs

- Boiling water-bath or beaker of hot water (from a freshly boiled kettle)
- Heat-proof mat
- 10 cm^3 measuring cylinder
- Syringe

SAFETY GUIDANCE

- Eye protection must be worn.
- Wear a lab coat at all times. Handle all solutions with care. Avoid contact with skin, eyes or clothing. If solutions come into contact with your eyes or skin, wash immediately.
- Iodine solution causes stains and may be an irritant [H] to the eyes.
- Copper sulphate solution (also contained in Benedict's solution) is harmful [H] if swallowed and can cause skin or eye irritation.
- Sodium hydroxide is an irritant [H].
- Alcohol (ethanol) is harmful [H] if swallowed and highly flammable [F]. Keep away from naked flames.

Test for starch

- Starch powder
- Iodine solution

Test for reducing sugar

- Glucose solution
- Benedict's solution

Test for protein

- 1% solution of albumin (egg white)
- Biuret solution (equal volume of 0.1 M sodium hydroxide solution and 0.01 M copper sulfate solution)

Test for fat

- Cooking oil
- Alcohol (ethanol)

Test for vitamin C

- Fresh lemon juice
- Fresh orange juice
- 0.1% DCPIP solution

Method

Test for starch

1 Place a spatula full of starch powder in a test tube.

2 Add 5 cm³ warm water to the test tube. Shake gently to make a suspension.

3 Add a few drops of iodine solution.

4 Record your observations.

Test for reducing sugar

1 Place 1 cm³ glucose solution in a test tube.

2 Add 1 cm³ of Benedict's solution to the same test tube.

3 Heat the test tube in a water-bath or a beaker of hot water (above 80 °C) for 5 minutes.

4 Record your observations.

Test for protein

1 Place 1 cm³ 1% solution of albumin in a test tube.

2 Add 1 cm³ biuret solution to the same test tube.

3 Mix well and record your observations.

Test for fat

1 Place 2 drops of cooking oil in a test tube.

2 Add 2 cm³ alcohol (ethanol) to the same test tube. Shake gently.

3 Add 2 cm³ of water. Shake gently.

4 Allow to stand for a few minutes before you record your observations.

Test for vitamin C

1 Place 2 cm³ 0.1% DCPIP solution in a test tube.

2 Draw up 2 cm³ fresh lemon juice into a syringe. Add the juice drop by drop to the test tube.

3 Record the volume of juice used to change the colour of DCPIP.

4 Repeat steps 1, 2 and 3 using orange juice.

TIP

- These are qualitative tests, which means that they tell you whether the food type is present or not. It does not tell you how much is present.

- Slight colour changes are often difficult to judge. Hold up the tube against a white background such as a piece of paper. A control (a second experiment set up with all the other conditions the same) with water instead of the sample will help to show slight changes in colour.

- Keep beakers of boiling water well away from you so that you don't knock them over. If you are sharing a water-bath with others, label your test tubes with your initials to avoid mixing yours up with other students'.

- It can be difficult to see the cloudy emulsion. Allow the test tube to stand for some time. You will see the emulsion at the top.

Observations

Starch test: ....

Benedict's test: ....

Biuret test: ....

Emulsion test: ....

DCPIP test – lemon juice: ....

DCPIP test – orange juice: ....

Conclusions

What can you conclude from the results of the DCPIP test on orange juice and lemon juice?

....

....

....

....

Evaluation

1   Other than usual safe practices in the laboratory, list the safety precautions to be taken when carrying out each of the above food tests. Give a reason for each precaution.

....

....

....

....

....

....

....

....

....

2   Suggest a control for the tests so that you can observe a negative result when none of the food type is present.

....

....

....

GOING FURTHER

• •

1 Suggest how you could make the vitamin C test quantitative (able to compare how much is present).

....

....

....

2 It is difficult to judge the colour change in some of the tests. Research how the method can be improved to make the readings easier to interpret.

....

....

4.2 Application of food tests

Foods contain a variety of food types, which can be identified using food tests. It is useful to know which of these are present in food so that we can eat a balanced diet.

KEY TERMS

Benedict's solution
biuret test
DCPIP test
iodine solution
reducing sugar

Aim

To test a range of food samples for the presence of starch, reducing sugar, protein, fat and vitamin C.

Apparatus

- Eye protection
- As for Experiment 4.1 'Food tests' on pages 38–39
- Food samples provided by your teacher, such as potato, apple, onion, raisins, beans, meat and sunflower seeds. Solids can be cut into small pieces
- Pestle and mortar
- Water
- Funnel
- Filter paper
- Droppers
- Alcohol

SAFETY GUIDANCE

- Eye protection must be worn.
- Do not eat the food substances provided for testing.
- Wear a lab coat at all times. Wear gloves and handle all solutions with care. Avoid contact with skin, eyes or clothing. If solutions come into contact with your eyes or skin, wash immediately.
- Iodine solution causes stains and may be an irritant to the eyes.
- Copper sulphate solution (also contained in Benedict's solution) is harmful [H] if swallowed and can cause skin or eye irritation.
- Sodium hydroxide is an irritant at this concentration.
- Alcohol (ethanol) is harmful [H] if swallowed and highly flammable [F]. Keep away from naked flames.

Method

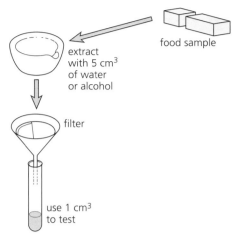

food sample

extract
with 5 cm³
of water
or alcohol

filter

use 1 cm³
to test

Figure 1

TIP

Make sure to clean the pestle and mortar each time. Contamination will give false results.

Some foods contain more than one food type.

1 Crush potato, apple, onion, raisins and beans **individually**, using the pestle and mortar, in about 5 cm³ water. Crush meat and sunflower seeds, using the pestle and mortar, in about 5 cm³ alcohol. Use 1 cm³ of the liquid to test for the presence of starch, reducing sugar, protein, fat and vitamin C.

2 Follow the instructions in Experiment 4.1 'Food tests' on page 39 for each test.

3 Record your observations in the table below.

Observations

Record the change of colour, if any. Also place a tick (✓) for presence, and a cross (×) for absence for each of the foods tested. If you are not sure, leave the box blank. The first one has been done as an example for you.

Table 1

Food tested	Starch test		Reducing sugar test		Protein test		Fat test		Vitamin C test	
	Final colour	✓/×	Final colour	✓/×	Final colour	✓/×	Final colour	✓/×	Final colour	✓/×
potato	blue/black	✓								

Conclusions

For each of the food samples you tested, list the food types that were present.

....

....

....

....

....

....

....

Evaluation

Describe any problems that you had in interpreting your observations.

....

....

....

GOING FURTHER

1   Why do you think it is important to know the food type of different foods?

    ....

    ....

2   Suggest how this experiment, or an improvement to this experiment, could help to compare the quantity of a food type in different foods.

    ....

    ....

    ....

    ....

5 Enzymes

5.1 Testing the effect of high temperature on an enzyme

Enzymes are proteins that function as biological catalysts, speeding up chemical reactions in cells. High temperatures denature enzymes and so prevent them from functioning. Liver cells contain the enzyme catalase, which breaks down hydrogen peroxide, a toxic by-product of many chemical reactions.

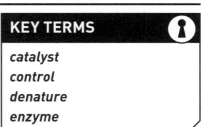

KEY TERMS

catalyst
control
denature
enzyme

Aim

To test the effect of high temperature on the enzyme catalase.

KEY EQUATION

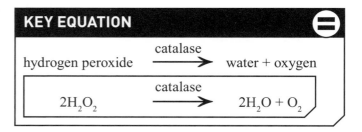

hydrogen peroxide $\xrightarrow{\text{catalase}}$ water + oxygen

$$2H_2O_2 \xrightarrow{\text{catalase}} 2H_2O + O_2$$

Apparatus

- Eye protection
- Fresh piece of liver
- Pestle and mortar, with sand
- 4 test tubes
- 2 dropping pipettes
- Filter funnel and paper
- Water
- 10 Vol/3% hydrogen peroxide solution
- Measuring cylinder
- Hot water-bath or beaker of hot water (from a freshly boiled kettle)

Method

1 Grind a small piece of liver with $20\,\text{cm}^3$ water and some sand using a pestle and mortar.

2 Filter the mixture into two test tubes (see Figure 1, page 42). Label them 'A' and 'B'.

3 Pipette a few drops of the filtrate from test tube A into a test tube containing $3\,\text{cm}^3$ 10 Vol/3% hydrogen peroxide solution. Record your observations.

4 Heat the filtrate in test tube B in a beaker of hot water for 2 minutes.

5 Add a few drops of filtrate from test tube B to another test tube containing $3\,\text{cm}^3$ 10 Vol/3% hydrogen peroxide solution. Record your observations.

6 Wash your hands thoroughly.

SAFETY GUIDANCE

- Eye protection must be worn.
- Wear a lab coat at all times. Handle all solutions with care. Avoid contact with skin, eyes or clothing. If solutions come into contact with your eyes or skin, wash immediately.
- Liver is a biological specimen. Dispose of it in the appropriate bin.
- Hydrogen peroxide is an irritant. It may cause burns to the skin and damage to the eyes. Take care not to touch your face during the practical in case your hand has hydrogen peroxide on it.

Observations

Record your observations below.

....

....

Conclusions

Explain what you have observed.

TIP

The equation for the reaction may help.

....

....

....

Evaluation

1   Explain why the liver was ground in the pestle and mortar.

....

2   Suggest a suitable control for this investigation.

....

3   Describe how you would modify this experiment to monitor the rate of reaction over time. Draw a suitable table for collecting results.

....

....

....

....

GOING FURTHER

• •

Normal human body temperature is about 37 °C. Body temperatures over 41 °C can damage organs and even cause death. Suggest an explanation why.

....

....

5.2   The effect of temperature on enzyme reaction

Enzymes are biological catalysts. They work best in certain conditions. Temperature will affect the activity of an enzyme, but high temperatures denature enzymes and so prevent them from functioning. The temperature at which an enzyme works best is called its optimum temperature.

Amylase is a digestive enzyme that breaks down starch into smaller molecules (glucose). We can test for the presence of starch by adding iodine solution. Iodine solution reacts with starch grains, giving a blue/black colour. If no starch is present, this means the substrate (starch) has broken down to glucose.

However, it is sometimes difficult to detect the end point of a reaction when observing subtle colour changes, especially if other coloured substances are present, in which case colorimeters can be used.

Aim

To investigate the effect of temperature on the rate of an enzyme reaction.

Apparatus

- Eye protection
- 5% amylase solution
- 1% starch solution
- Dilute iodine solution
- Plastic syringes or graduated pipettes (1 cm³ and 5 cm³)
- Dropping pipette
- 6 test tubes
- 3 beakers
- Thermometer
- Stopwatch
- Sources of warm, cold and ice water

KEY TERMS

catalyst
denature
enzyme
optimum temperature

SAFETY GUIDANCE ⚠

- Eye protection must be worn.
- Wear a lab coat at all times. Handle all solutions with care. Avoid contact with skin, eyes or clothing. If solutions come into contact with your eyes or skin, wash immediately.
- Enzymes can cause allergies. Take care while handling them [H].
- Avoid using temperatures higher than 50 °C, as there is a risk of scalding.

Photocopying prohibited

Cambridge IGCSE™ Biology Practical Skills Workbook

Method

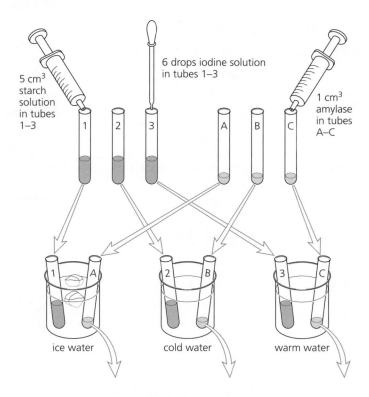

6 drops iodine solution in tubes 1–3

5 cm³ starch solution in tubes 1–3

1 cm³ amylase in tubes A–C

ice water   cold water   warm water

leave all three for 5 minutes

ice water   cold water   warm water

add the amylase to the starch solution and start the stopwatch

Figure 1

1 Label three test tubes 'A', 'B' and 'C'.

2 Label another three test tubes '1', '2' and '3'.

3 Using a syringe (or graduated pipette), place 1 cm³ of 5% amylase solution into test tubes A, B and C.

4 With a clean syringe, place 5 cm³ of 1% starch solution into test tubes 1, 2 and 3.

5 Using the dropping pipette, add six drops of the dilute iodine solution to test tubes 1, 2 and 3. Record your observations.

6 Prepare three beakers with water temperatures of about 10 °C, 20 °C and 35 °C.

7 Place the tubes in the beakers so that
   – the ice water contains test tubes 1 and A
   – the cold water contains test tubes 2 and B
   – the warm water contains test tubes 3 and C.

> **TIP**
>
> Rinse pipettes and syringes several times with clean distilled water to prevent contamination.

> **TIP**
>
> A mix of ice in water will give a water temperature of about 10 °C. Water from the cold tap should be at about 20 °C. Obtain warm water at about 35 °C by mixing warm and cold water.

8 Leave for 5 minutes so that the test tubes reach the temperature of the surrounding water.

9 After 5 minutes record the temperature of each of the beakers.

10 Pour the contents of tube A into tube 1, tube B into tube 2 and tube C into tube 3.

11 Start the stopwatch and record the time it takes to observe a colour change in each of the tubes.

> **TIP**
>
> You may need to keep adding ice during the experiment to keep the temperature of the ice and water beaker at about 10 °C. Similarly, add a small amount of hot water to keep the temperature of the warm water beaker at about 35 °C.

Observations

Record the colour change observed after adding iodine to starch solution in test tubes 1, 2 and 3 in the table below.

Table 1

	Temperature/°C	Time taken for colour change/s
Tube 1		
Tube 2		
Tube 3		

Conclusions

1 Complete the following sentences:

  a The enzyme .... catalyses the breakdown of .... into smaller molecules.

  b In this investigation, the colour of the iodine solution at the beginning was ....

   When all the starch was broken down, the iodine solution was ....

2 What pattern (trend) did you observe?

   ....

3 At what temperature was the enzyme most active? How did you know?

   ....

> **TIP**
>
> The rate of reaction is the time it takes to change the substrate (starch) into product (glucose).

4 How did temperature affect the rate of reaction?

   ....

5 Explain the trend you observed.

   ....

   ....

   ....

   ....

Evaluation

Outline how this experiment could be improved to make a more accurate estimate of the optimum temperature for this enzyme.

....

....

....

....

5.3   The effect of pH on enzyme reaction

Enzymes are biological catalysts which work best within specific optimum conditions. The level of acidity or alkalinity (pH) will affect enzyme activity. Many enzymes are denatured at very high or very low pH. The pH at which an enzyme works best is called its optimum pH.

Amylase is an enzyme produced in saliva in the mouth and in the small intestine. Amylase breaks down starch into sugars. The iodine test confirms the presence of starch.

Aim

To investigate the effect of pH on the rate of an enzyme reaction.

Apparatus

- Eye protection
- 5 test tubes
- Plastic syringes
- Dropping pipette
- 1% starch solution
- 0.05 M sodium carbonate solution
- 0.1 M ethanoic (acetic) acid
- Dilute iodine solution
- 5% amylase solution
- Cavity tile
- Beaker of water
- Stopwatch
- pH paper and colour chart of pH values

KEY TERMS

catalyst
denature
enzyme
optimum pH

SAFETY GUIDANCE

- Eye protection must be worn.
- Wear a lab coat at all times. Handle all solutions with care. Avoid contact with skin, eyes or clothing. If solutions come into contact with your eyes or skin, wash immediately.
- Sodium carbonate solution is a low hazard at this concentration.
- Ethanoic (acetic) acid is an irritant [H] to the skin and eyes.
- Enzymes can cause allergies. Take care while handling them [H].
- Iodine solution causes stains and may be an irritant [H] to the eyes.

Method

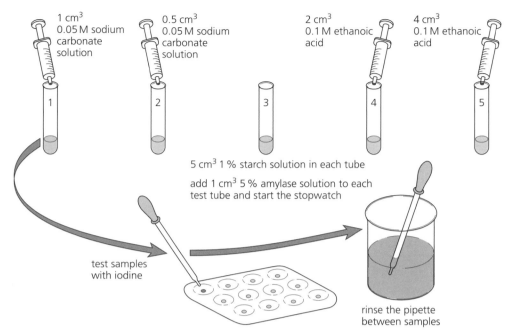

Figure 1

1 Label the test tubes 1–5.

2 Using the syringe, place 5 cm³ of 1% starch solution in each test tube.

3 Adjust the pH of the starch solution by adding the amounts of acid or alkali shown in Figure 1 and Table 1 to the appropriate tube. Use a clean syringe when changing from sodium carbonate solution to ethanoic acid.

Table 1

Tube	Chemical		Approximate pH
1	1 cm³ 0.05 M sodium carbonate solution	9	alkaline
2	0.5 cm³ 0.05 M sodium carbonate solution	7–8	slightly alkaline
3	nothing	7	neutral
4	2 cm³ 0.1 M ethanoic (acetic) acid	6	slightly acidic
5	4 cm³ 0.1 M ethanoic (acetic) acid	3	acidic

4 Check the pH of the solutions by placing a drop from each tube onto pH paper and comparing it with a colour chart of pH values. Record the pH of the solutions.

5 Using a pipette, place a drop of 0.01 M iodine solution in each cavity of the tile.

6 Place 1 cm³ of 5% amylase solution into tube 1.

7 Gently shake the test tube to mix the contents. Take care to avoid spills.

8 Using a clean dropping pipette, quickly remove a small sample from tube 1 and let one drop fall onto one of the iodine drops on the cavity tile. Start the stopwatch immediately and record the colour of the iodine. Return the remainder of the solution in the pipette to tube 1.

9 After 30 seconds, take another sample from tube 1 and repeat the iodine test. Do this every 30 seconds for 5 minutes, or until the iodine solution remains orange.

10 When the sample no longer gives a blue colour, record the time.

11 Rinse the pipette in a beaker of water.

12 Repeat steps 6 to 11 with the next tube.

> **TIP**
>
> Always rinse the pipette well between samples to avoid contamination.

Observations

1 Record your observations in the table below.

Table 2

Time/s	Tube 1 pH	Tube 2 pH	Tube 3 pH	Tube 4 pH	Tube 5 pH .....
	Colour of iodine solution				
0					
30					
60					
90					
120					
150					
180					
210					
240					
270					
300					

2 Plot a bar chart of your results. Place approximate pH on the x-axis and the time (in appropriate units) on the y-axis.

TIP

Remember the 'nappy' rule when plotting variables on a graph: what you can change goes on the bottom.

Conclusions

1 In which test tube was the enzyme (amylase) most active? How did you know?

....

....

2 How did pH affect enzyme activity?

....

....

3 What do you think is the optimum pH for this enzyme? Explain why.

....

....

Evaluation

1 Why was it necessary to rinse the pipette before repeating the procedure with a different tube?

....

2 Complete the following sentence.

It was important to note the time at which the samples no longer produced a blue colour with iodine solution because this shows that all of the starch in that sample had been broken down into .... by the enzyme ....

3 What was the independent variable being tested in this experiment?

....

4 Describe how the experiment could be improved to make sure that this is the only variable affecting the results.

....

....

5 Describe how you could obtain a more accurate estimate of the optimum pH for this enzyme.

....

....

....

....

6 Plant nutrition

6.1 The importance of different mineral ions for plant growth

Like all living things, plants need certain mineral ions to grow well. They get these from the soil by absorption through the roots. We can see what effect these ions have on plant growth by growing seedlings in different conditions. It takes time for seeds to germinate and grow into seedlings. We can only observe the effect of the conditions when the leaves have emerged.

Aim

To investigate the importance of different mineral ions for plant growth.

Apparatus

- Eye protection
- 4 test tubes
- 15 wheat seeds
- 4 different growth mediums: (A) complete growth medium; (B) no nitrates; (C) no magnesium and (D) distilled water
- Aluminium foil
- Cotton wool
- Distilled water

KEY TERMS

amino acid
chlorophyll
germinate

SAFETY GUIDANCE

- Eye protection must be worn.
- This practical presents minimal risk. However, standard laboratory safety rules still apply at all times.
- Treat glassware with care to avoid breakages and cuts.

Method

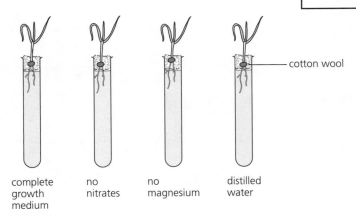

complete growth medium | no nitrates | no magnesium | distilled water — cotton wool

Figure 1

1 Label the test tubes A, B, C and D.

2 Set up the experiment as shown in Figure 1, placing three wheat seedlings in each of the test tubes containing the different growth mediums.

3 Cover the test tubes with foil.

4 Leave the seedlings to grow in these solutions for 3 weeks, making sure you keep the tubes topped up with distilled water.

5 Record your observations below.

Observations

Use the table below to record your observations and conclusions.

Table 1

Growth medium		Observations	Explanations
A	complete growth medium		
B	no nitrates		
C	no magnesium		
D	distilled water		

Evaluation

1 Explain why the test tubes should be topped up with distilled water and not tap-water.

....

....

2 Suggest why the test tubes were covered in foil.

....

3 Outline how this experiment could be improved to make it more reliable.

....

....

....

....

GOING FURTHER

Hydroponics is a technique of growing plants without soil. Undertake some research to find out about this technique and what type of plants are often grown using hydroponics.

1 With your understanding of photosynthesis and plants' requirements, plan out how you would grow plants this way to get a good yield.

....

....

....

....

....

2 Suggest what the nutrient medium should contain.

....

....

6.2   Is chlorophyll necessary for photosynthesis?

Most plants have green leaves because they contain a green pigment called chlorophyll, which helps them capture light energy from the Sun to photosynthesise. Some plants have variegated leaves (leaves that have non-green parts). Only the green part of the leaf is able to photosynthesise and make glucose. Excess glucose made by photosynthesis is quickly converted to starch. We can use the iodine test to show the presence of starch in leaves that have been photosynthesising.

Aim

To investigate whether chlorophyll is necessary for photosynthesis.

Apparatus

- Eye protection
- Plant with variegated leaves (e.g. a geranium)
- Heat-proof mat or tile
- Hot water-bath or hot water from a kettle
- 2 glass beakers
- Test tube
- Forceps
- Alcohol (ethanol)
- White tile
- Iodine solution
- Stopwatch

KEY TERMS

chlorophyll
photosynthesis

SAFETY GUIDANCE

- Eye protection must be worn.
- Iodine solution causes stains and may be an irritant [H] to the eyes. Take care to avoid getting it on skin, clothes or equipment. If it comes into contact with your eyes or skin, wash immediately.
- Alcohol (ethanol) is harmful [H] if swallowed and highly flammable [F]. Keep away from naked flames.

Method

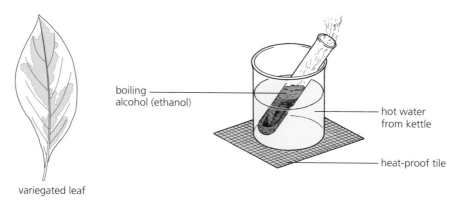

boiling alcohol (ethanol)

hot water from kettle

heat-proof tile

variegated leaf

Figure 1

1 Carefully remove a variegated leaf from a plant.

2 Draw a diagram of this leaf, recording areas where there is chlorophyll (green) and areas where there is none.

3 Remove the chlorophyll from the leaf by using the following method:
   a Half-fill a beaker with hot water from a kettle.
   b Dip the leaf in the hot water, using forceps, and leave for 30 seconds.
   c Place the leaf in a test tube and push it to the bottom.
   d Cover the leaf with alcohol (ethanol) and place the test tube in the hot water-bath or beaker of hot water (see Figure 1).
   e Ethanol boils at 78°C. It will boil in hot water and dissolve the chlorophyll.
   f Pour the ethanol into a spare beaker.
   g Using the forceps, carefully remove the decolourised leaf from the test tube and flatten it out onto a white tile.

4 Add a few drops of iodine solution to the leaf.

5 Record your observations and draw a labelled diagram of your observations of the leaf.

> **TIP**
>
> The leaf is placed into hot water to destroy the cell membranes. This makes the leaf more permeable so chlorophyll can come out of the cells and iodine solution can get into the cells.

> **TIP**
>
> Take care when removing the leaf from the alcohol as it becomes brittle and is easily torn.

Observations

Diagram of leaf before test

Diagram of leaf after test

Summarise your observations.

....

....

Conclusions

Explain your observations.

....

....

....

....

Evaluation

1   If the leaf was not submerged in hot water, what problem would this cause with the method?

....

....

2   Explain why it is important to remove chlorophyll from the leaf by dissolving it in alcohol.

....

....

....

3   Outline how this experiment could be improved to make it more reliable.

....

....

....

GOING FURTHER

1   If you observe different plants, you will see that their leaves are different shades of green. Some leaves are even red, orange and purple. Undertake some research to find out why leaves are different colours and how a red-leaved plant survives.

....

....

....

....

> ....
>
> ....
>
> **2** Predict how the leaves of a red-leaved plant would look if you carried out the same experiment as described above. Give a reason for your prediction.
>
> ....
>
> ....
>
> ....

6.3   Is light necessary for photosynthesis?

You will observe from the photosynthesis equation that plants require light, carbon dioxide and water to photosynthesise and produce glucose. The excess glucose is stored as starch.

KEY EQUATION ☰

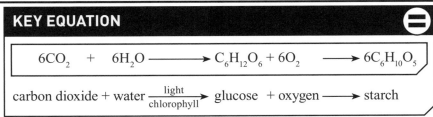

$$6CO_2 \quad + \quad 6H_2O \longrightarrow C_6H_{12}O_6 + 6O_2 \longrightarrow 6C_6H_{10}O_5$$

carbon dioxide + water $\xrightarrow[\text{chlorophyll}]{\text{light}}$ glucose + oxygen $\longrightarrow$ starch

To investigate whether the presence of light affects starch production, we have to keep a plant in the dark for about 48 hours so that all the existing starch is used up. This is called destarching.

Aim

To investigate whether light is necessary for photosynthesis.

Apparatus

- Eye protection
- Plant with green leaves that has been left in darkness for a few days, so the leaves become destarched
- Aluminium foil
- Heat-proof tile
- Beaker of hot water from a kettle
- Scissors
- Test tube
- Alcohol (ethanol)
- Iodine solution

KEY TERMS

chlorophyll
control
destarch
photosynthesis

SAFETY GUIDANCE

- Eye protection must be worn.
- Iodine solution causes stains and may be an irritant [H] to the eyes. Take care to avoid getting it on skin, clothes or equipment. If it comes into contact with your eyes or skin, wash immediately.
- Alcohol (ethanol) is harmful [H] if swallowed and highly flammable [F]. Keep away from naked flames.

Method

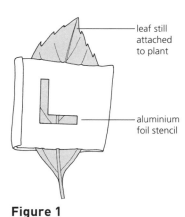

leaf still attached to plant

aluminium foil stencil

Figure 1

1 Remove the chlorophyll from one leaf and test the leaf for the presence of starch (see Experiment 6.2 'Is chlorophyll necessary for photosynthesis?' on page 55).

2 Record your observations.

3 Cut a shape into a piece of aluminium foil.

4 Attach the foil shape to a leaf that is still on the destarched plant.

5 Position the plant in sunlight and leave for about 6 hours.

6 Remove the leaf, then remove the chlorophyll from the leaf and test it for starch.

7 Record your observations and make a labelled drawing.

> **TIP**
>
> In any investigation there should only be one variable altered: the one you are investigating. All other variables must be kept constant (controlled).

Observations

1 What do you observe after conducting the test in step 1?

....

....

2 What do you observe after conducting the test in step 6?

....

....

3 Make a labelled drawing of your observations below.

Conclusions

Explain your observations.

....

....

....

....

Evaluation

1 Why is aluminium foil used?

....

2 Why was it important to use a destarched plant for this investigation?

....

....

3 What was the purpose of testing a leaf for starch at the start?

....

....

4 Other than placing the plant in a dark cupboard for a few days, how else could the leaf have been destarched?

....

....

....

5 Outline how this experiment could be improved by using a control experiment to show that the effect you have observed is not due to the aluminium foil preventing carbon dioxide from entering the leaf.

....

....

....

GOING FURTHER

White light is a mixture of different colours. You can see these colours in a rainbow or if you pass white light through a glass prism. The colour of an object depends on the colour it reflects. Green plants appear green because they reflect green light.

Design an experiment to see which colour of light is most effective for photosynthesis.

....

....

....

....

6.4   Is carbon dioxide necessary for photosynthesis?

According to the equation for photosynthesis, carbon dioxide is a necessary substrate. We can use the iodine test to show the presence of starch in leaves that have been photosynthesising.

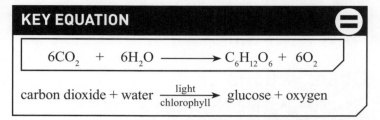

KEY EQUATION

$$6CO_2 \quad + \quad 6H_2O \quad \longrightarrow \quad C_6H_{12}O_6 + \quad 6O_2$$

$$\text{carbon dioxide} + \text{water} \xrightarrow[\text{chlorophyll}]{\text{light}} \text{glucose} + \text{oxygen}$$

Aim

To investigate whether carbon dioxide is necessary for photosynthesis.

Apparatus

- Eye protection
- 2 destarched green plants (kept in the dark for 48 hours)
- 2 plastic bags
- 2 elastic bands
- Soda lime and dilute sodium hydrogencarbonate solution
- Iodine solution
- Heat-proof tile
- Hot water-bath or hot water from a kettle
- 2 glass beakers
- Test tube
- Forceps
- Alcohol (ethanol)
- White tile

Method

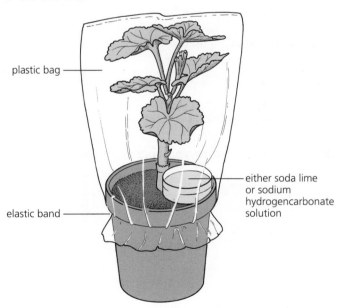

plastic bag

either soda lime
or sodium
hydrogencarbonate
solution

elastic band

Figure 1

1   Place a pot of soda lime onto one of the plant pots and a pot of sodium hydrogencarbonate solution onto the other plant pot.

2   Cover both plants with clear plastic bags and secure with elastic bands.

KEY TERMS

destarch

photosynthesis

TIP

When sodium hydrogencarbonate is dissolved in water it releases carbon dioxide into the air. Soda lime absorbs carbon dioxide from the air.

SAFETY GUIDANCE

- Eye protection must be worn.
- Wear a lab coat at all times. Handle all solutions with care. Avoid contact with skin, eyes or clothing. If solutions come into contact with your eyes or skin, wash immediately.
- Soda lime is corrosive [C] and irritant [H] to eyes and skin. **Only the teacher or technician should handle the soda lime.**
- Sodium hydrogencarbonate solution is low hazard.
- Iodine solution causes stains and may be an irritant [H] to the eyes.
- Alcohol (ethanol) is harmful [H] if swallowed and highly flammable [F]. Keep away from naked flames.

3 Place both plants in the light for several hours.

4 Remove a leaf from each plant, then remove the chlorophyll from each leaf and test them for the presence of starch (see Experiment 6.2 'Is chlorophyll necessary for photosynthesis?' on page 55).

5 Record your observations.

Observations

Observation of leaf after adding iodine solution to
- the plant with soda lime:

....

- the plant with sodium hydrogencarbonate solution:

....

Conclusions

How can you explain what you observed?

....

....

....

....

Evaluation

1 Why were the plants kept in the dark for two days before the experiment?

....

....

2 Explain why soda lime was added to one of the plants and sodium hydrogencarbonate to the other.

....

....

....

....

3 Suggest why the plants were each sealed in a plastic bag with an elastic band.

....

4 Suggest how this experiment could be improved to make it more reliable.

....

....

....

5 Plan how you would carry out an investigation to test the effect of different concentrations of carbon dioxide on the rate of photosynthesis.

Hint: to measure the rate of photosynthesis, you will need to compare how quickly each plant is photosynthesising. How could you do this?

....

....

....

....

....

....

....

6 Suggest a control for your planned investigation.

....

GOING FURTHER
• •

Growing plants in a greenhouse allows growers to alter the carbon dioxide levels in the greenhouse by pumping in gas. Suggest how you would monitor the effect of changes in the carbon dioxide levels on plant growth.

....

....

....

....

6.5 Gas exchange during photosynthesis

During the day, plants take in carbon dioxide from the air and combine it with water to make glucose by the process of photosynthesis. The waste gas they produce is oxygen. Some of the glucose is used for energy, and the excess is stored as starch. However, to release the energy from glucose they need to respire, and for this they need to take in oxygen. The process of respiration releases carbon dioxide.

At night, plants stop photosynthesising but continue to respire, breaking down glucose and releasing carbon dioxide. The gases enter and leave the leaves through tiny pores on the leaf surface called stomata.

Aim

To investigate gas exchange during photosynthesis.

Apparatus

- Eye protection
- 3 test tubes with bungs
- Test-tube rack
- Dilute hydrogencarbonate indicator
- Aquatic plant such as pondweed, *Elodea*
- Aluminium foil
- Distilled water
- Stopwatch
- Measuring cylinder

Method

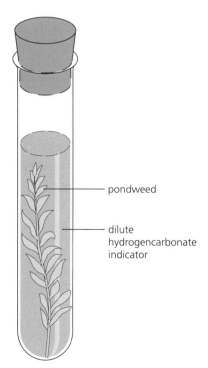

pondweed

dilute hydrogencarbonate indicator

Figure 1

KEY TERMS

gas exchange
photosynthesis
respiration
stoma (plural: stomata)

SAFETY GUIDANCE

- Eye protection must be worn.
- Wear a lab coat at all times. Handle all solutions with care. Avoid contact with skin, eyes or clothing. If solutions come into contact with your eyes or skin, wash immediately.
- Sodium hydrogen-carbonate solution is a low hazard.
- Wash your hands thoroughly after handling pondweed.

TIP

Hydrogencarbonate indicator is used to indicate the levels of dissolved carbon dioxide by changes in colour. The colour changes are:

Table 1

Colour	Carbon dioxide level
yellow	very high
orange	high
red	normal colour for atmospheric carbon dioxide
magenta	low
purple	very low

1 Rinse the inside of three test tubes with tap-water, then distilled water, then dilute hydrogencarbonate indicator.

2 Place 5 cm³ of hydrogencarbonate indicator into each of the tubes.

3 Label the tubes 1–3.

4 Place a piece of pondweed in tubes 1 and 2.

5 Place bungs into all of the tubes.

6 Cover tube 1 with aluminium foil.

7 Observe the colour of the hydrogencarbonate indicator, then place all three tubes in sunlight.

8 Leave for 40 minutes.

9 Record your observations in Table 2.

Observations

Table 2

Tube	Colour of indicator at start	Colour of indicator after 40 minutes
1		
2		
3		

Conclusions

Explain your observations.

....

....

....

....

Evaluation

1 Why were the test tubes rinsed in step 1?

....

2 Suggest at least one reason why this experiment may not be very reliable. Outline an improvement you could make.

....

....

....

....

....

GOING FURTHER

Design an experiment to investigate how much oxygen is produced by photosynthesis per hour by an aquatic plant in full sunlight. Outline one problem with this investigation, and explain how the problem might affect the results.

....

....

....

....

....

....

....

6.6 The effect of light intensity on the rate of photosynthesis

Light is essential for photosynthesis. An increase in light intensity may therefore affect the rate of photosynthesis. The rate of photosynthesis for a piece of pondweed in different conditions can be compared by counting the number of bubbles of oxygen produced over 1 minute. Light intensity can be varied by moving a lamp closer to the plant. Light intensity is inversely proportional to distance from the light source.

Aim

To investigate the effect of light intensity on the rate of photosynthesis.

Apparatus

- Eye protection
- Beaker of water
- Saturated sodium hydrogencarbonate solution
- Canadian pondweed or *Cabomba caroliana* (ensure this is sourced and disposed of appropriately)
- Scalpel or knife
- Paper clip
- Bench lamp
- Metre rule
- Stopwatch
- Measuring cylinder

KEY TERMS

light intensity
photosynthesis

SAFETY GUIDANCE

- Eye protection must be worn.
- Wear a lab coat at all times. Handle all solutions with care. Avoid contact with skin, eyes or clothing. If solutions come into contact with your eyes or skin, wash immediately.
- Sodium hydrogen-carbonate solution is a low hazard.
- Keep the electric lamp away from water and do not touch the power pack, plug or socket with wet hands.
- Take care to avoid cutting yourself with the scalpel/knife. Always cut directionally away from yourself.
- Wash your hands thoroughly after handling pondweed.

Method

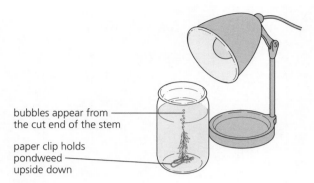

bubbles appear from the cut end of the stem

paper clip holds pondweed upside down

Figure 1

1 Refer to Figure 1 and read the whole method before you begin.

2 Draw a suitable table to record your results.

3 Add 5 cm³ of saturated sodium hydrogencarbonate solution to a beaker of water.

4 Carefully cut a piece of pondweed or *Cabomba caroliana* shoot to about 10 cm in length.

5 Attach a paper clip to the tip of the shoot and place in the beaker of water.

6 Place a bench lamp close to the beaker and measure the distance between them.

7 Switch on the lamp. Allow the plant to acclimatise to the conditions for a few minutes, until bubbles start to emerge from the pondweed. (If no bubbles appear after 5 minutes, speak to your teacher.)

8 Start the stopwatch.

9 Count the number of bubbles produced over 1 minute.

10 Record your observations in your table.

11 Repeat steps 6–10, moving the lamp steadily further away; for example, 10 cm further away each time. You should aim to have five readings.

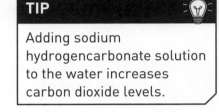

TIP

Adding sodium hydrogencarbonate solution to the water increases carbon dioxide levels.

Observations

1 Draw a table in which to record your results here.

2 Add a column to your table for $\frac{1}{d^2}$, where d = distance from the lamp. Calculate this value for each measurement you made.

3 Plot a line graph of mean value of bubbles collected against $\frac{1}{d^2}$ from the lamp.

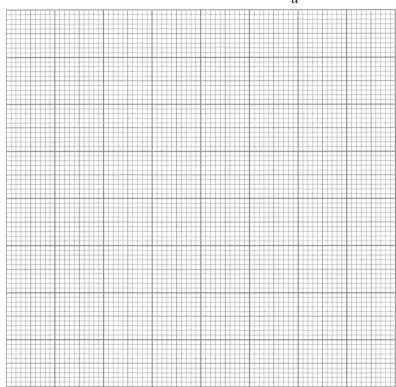

Conclusions

How did light intensity affect the rate of photosynthesis? Explain your answer.

....

....

....

....

Evaluation

1 Explain why sodium hydrogencarbonate solution was added to the beaker of water.

....

....

....

....

2 In any investigation all variables must be kept constant, except the one you are investigating. Were you able to control all the variables in this investigation? If not, suggest how you could have solved the problem.

....

....

....

....

....

....

3 Suggest what other problems there might be with the method used in this investigation, and how they could be overcome.

....

....

....

....

4 Design an investigation to observe the effect of temperature on the rate of photosynthesis.

....

....

....

....

GOING FURTHER
• •

High concentrations of nitrate and phosphate in a pond make algae and plants grow faster. This is called eutrophication. When too much algae grows on the surface of a pond, the algae prevent light reaching the other plants beneath. The faster the algae grow, the more oxygen they take out of the water. Both these factors can kill plants because they cannot photosynthesise, and can also kill animals living in the water.

Scientists investigating eutrophication can use a light meter to measure light levels in the water and can also use an oxygen meter to measure oxygen concentration in the water.

Design an investigation to measure how increased algal growth in a pond affects the organisms living in the pond.

....

....

....

....

7 Transport in plants

7.1 Transport in vascular bundles

Vascular bundles are specialised structures in plants that transport nutrients and water to all cells. Vascular bundles contain xylem and phloem tissue. The xylem transports water and mineral ions from the soil to all parts of the plant. The phloem transports food from the leaves or stem to wherever it is needed.

Aim

To investigate the transport of water in vascular bundles.

Apparatus

- Eye protection
- Celery stalk, with leaves, soaking in water
- Celery stalk, with leaves, soaking in 1% methylene blue solution
- Scalpel or knife
- Tile or cutting board
- Microscope
- Hand lens
- Microscope slides and cover-slips
- Mounted needle

Method

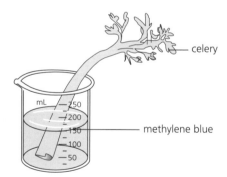

Figure 1

KEY TERMS

phloem
vascular bundles
xylem

SAFETY GUIDANCE

- Eye protection must be worn.
- Wear a lab coat at all times. Handle all solutions with care. Avoid contact with skin, eyes or clothing. If solutions come into contact with your eyes or skin, wash immediately.
- Methylene blue causes stains and may be an irritant [H] to the skin and eyes.
- Take care to avoid cutting yourself with the scalpel/knife. Always cut directionally away from yourself.

1 Your teacher will provide you with two celery stalks. One has been soaking in methylene blue dye, the other in water.

2 Place the celery stick soaked in methylene blue on the tile and cut thin sections across the stem (transversely). Try and find the thinnest section.

3 Use the hand lens to observe the section. If a microscope is available, observe and record your observations using the microscope.

4 Place a drop of water on a microscope slide.

5 Place the stem section on the slide and use the mounted needle to gently lower a cover-slip on top.

TIP

Look back at Experiment 2.1 'Looking at plant cells' on page 20 to help you observe using a microscope and make drawings of what you see.

6 Locate a vascular bundle using the microscope. Record your observations, under low and high power of your microscope, by making a large, labelled drawing.

7 Repeat steps 2–6 using the celery stick soaked in water.

8 Wash your hands thoroughly.

Observations

Stalk placed in dye   Stalk placed in water

Observe and record how far the dye has travelled up the stalk.

....

Any other observations: ....

....

....

Conclusions

1 Why do the stalks look different?

....

....

2 Can you tell from your observations which part of the vascular bundle contains the xylem and which part contains the phloem? Explain your answer.

....

....

....

....

Evaluation

Outline how this experiment could be improved to make it more reliable.

....

....

....

GOING FURTHER

• •

Undertake some research to find out how you could observe phloem in the vascular bundle.

....

....

....

7.2 Rates of water uptake in different conditions

Transpiration is the loss of water vapour from leaves. Anything that will increase evaporation of water from the surface of the leaves will increase the rate of transpiration. Water lost by evaporation is replaced by taking up more water through the stem, from the soil. The rate of uptake of water in a plant shoot is affected by a number of different environmental factors.

Aim

To investigate the rate of water uptake in a plant shoot under different conditions.

Apparatus

- Eye protection
- Leafy shoot
- Potometer apparatus (see Figure 1)
- 2 beakers
- Water
- Stopwatch
- Table fan or hair-dryer on cold setting
- Ruler

KEY TERMS

potometer
transpiration

SAFETY GUIDANCE

- Eye protection must be worn.
- Capillary tubes are fragile. Treat glassware with care to avoid breakages and cuts.
- Keep electrical appliances away from water and do not touch the power pack, plug or socket with wet hands.
- Wash your hands thoroughly after handling plants.

Method

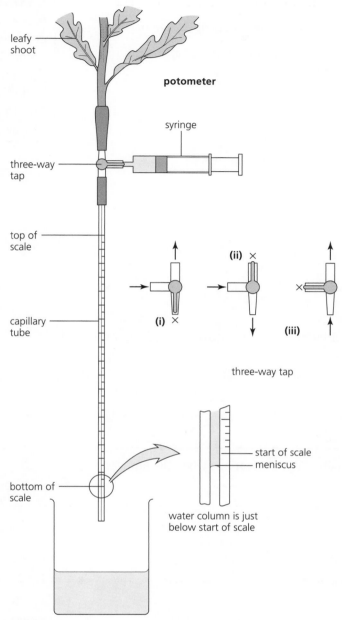

Figure 1

1 Keep the leafy shoot in water until you are ready to attach it to your potometer.

2 Without adding the leafy shoot, set up the potometer as shown in Figure 1, in a part of the room that is in direct sunlight, if possible.

3 Fill the syringe with water. Turn the three-way tap downwards and push the syringe until water comes out of the rubber tubing at the top.

4 Now gently push the leafy shoot into the rubber tubing as far as possible.

5 Turn the three-way tap upwards and press the syringe so that water comes out of the bottom of the capillary tube. Then turn the tap horizontally.

6 As the leaves transpire, water will be drawn from the capillary tube and the meniscus (the curved surface of the water in the tube) will rise.

7 Start your stopwatch. Record the distance moved by the meniscus after 1 minute.

TIP

If there is no movement of water in the capillary tube, remove the plant stem. Cut off about an inch from the end and quickly fit it back on the potometer.

8   Turn the tap upwards and, using the syringe, push the water to the bottom of the capillary tube. Turn the tap horizontally again.

9   Place your potometer by the table fan and turn the fan on.

10 Restart your stopwatch. Watch the meniscus rise again and record the distance moved after 1 minute.

> **TIP**
>
> Make sure that all the other conditions are kept the same. In any experiment there should be just one variable that you alter. All other conditions must remain the same.

Observations

Record your observations in a table below.

Conclusions

1   Explain how your observations demonstrate the process of transpiration.

....

....

2   Explain the difference in your observations when the fan was turned on.

....

....

....

....

....

Evaluation

1   Outline **two** other environmental conditions that could be changed to investigate their effect on rate of water uptake. Outline how you could change the method above to investigate each of these factors.

....

....

....

....

2 Explain why each condition could affect the rate of water uptake, and what effect it would have.

....

....

....

....

....

3 Suggest an explanation for the 'Tip' to cut off an inch of the stem if no movement of water is observed.

....

4 Outline how this experiment could be improved or made more reliable.

....

....

....

7.3 Which surface of a leaf loses more water vapour?

Plants lose water from their leaves by diffusion through tiny pores called stomata. Water vapour diffuses from a high concentration inside the leaf into the atmosphere, which has a lower concentration.

In this investigation we will observe whether there is a difference between the amount of water lost from the upper and lower surfaces of a leaf.

Aim

To investigate which side of a leaf loses more water vapour.

Apparatus

- 4 leaves (as close to the same size as possible; not from an evergreen plant)
- Vaseline® (or petroleum jelly)
- Newspaper (to protect bench)
- 2 retort stands
- Cotton thread

KEY TERMS

diffusion
stoma (plural: stomata)

SAFETY GUIDANCE

- This practical presents minimal risk. However, standard laboratory safety rules still apply at all times.
- Wash your hands thoroughly after handling plants.

Method

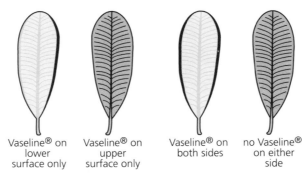

Vaseline® on lower surface only   Vaseline® on upper surface only   Vaseline® on both sides   no Vaseline® on either side

Figure 1

1   Use a permanent marker to number 4 leaves as 1, 2, 3 and 4.

2   Treat each of the leaves so that:
 – leaf 1 has Vaseline® on the lower surface only
 – leaf 2 has Vaseline® on the upper surface only
 – leaf 3 has Vaseline® on both sides
 – leaf 4 has no Vaseline® on either side.

3   Add Vaseline® to the end of each leaf stalk.

4   Tie a piece of cotton thread between two retort stands.

5   Suspend the leaves from the cotton thread for a few days.

6   Record your observations.

Observations

Record your observations below.

Leaf 1: ....

Leaf 2: ....

Leaf 3: ....

Leaf 4: ....

Conclusions

1   Which leaf lost the most water? How do you know?

 ....

2   Which loses more water – the lower or the upper surface of a leaf? Describe the evidence to support your conclusion.

 ....

 ....

 ....

3   Explain your observations for each leaf using ideas about the number of stomata.

 Leaf 1: ....

 ....

Leaf 2: ....

....

Leaf 3: ....

....

Leaf 4: ....

....

Evaluation

1 Explain why Vaseline® was added to the cut end of each leaf stalk.

....

2 Explain why it was important to use leaves of similar size.

....

....

3 Outline how this experiment could be improved to measure water loss more accurately.

....

....

....

....

....

GOING FURTHER

Do you think the water loss would be the same for all plants? Explain your answer.

....

....

....

Transport in animals

8.1   Investigating the effect of exercise on pulse rate

We can record pulse rate (number of beats per minute) as an indication of heart rate before and after exercise.

Aim

To investigate the effect of physical activity on pulse rate.

Apparatus

- Aerobics bench or step
- Stopwatch

Method

Measuring resting pulse rate

1   Practise measuring your pulse rate by finding your pulse in your wrist or neck.

2   Count the number of beats over 15 seconds, then multiply the result by 4. This is your resting pulse rate, measured in beats per minute.

3   Outline the essential safety procedures required for this activity. Check these with your teacher before you begin.

....

....

....

....

KEY TERM

pulse rate

SAFETY GUIDANCE

- This practical presents minimal risk. However, standard laboratory safety rules still apply at all times.
- If you have any health conditions, such as asthma, tell your teacher and follow advice.

Effect of exercise on pulse rate

1   Sit still and measure your resting pulse rate as described above. Record this below.

2   You are going to exercise for different periods of time and investigate how this affects your pulse rate immediately after exercise. Draw a table in the Observations section to record your results.

3   Exercise for 1 minute by stepping up and down on the step or by jumping on the spot. Measure your pulse rate again afterwards and record your findings.

4   Allow your pulse rate to return to normal (your resting pulse rate).

5   Repeat this activity for 2, 3, 4 and 5 minutes of exercise, each time allowing your pulse to return to your resting pulse rate before starting again.

Observations

1 Resting pulse: ....

2 Draw a table to record your results below.

3 Plot a graph with time spent exercising on the *x*-axis and pulse rate immediately after exercise on the *y*-axis. Plot the reading for your resting pulse at 0 minutes.

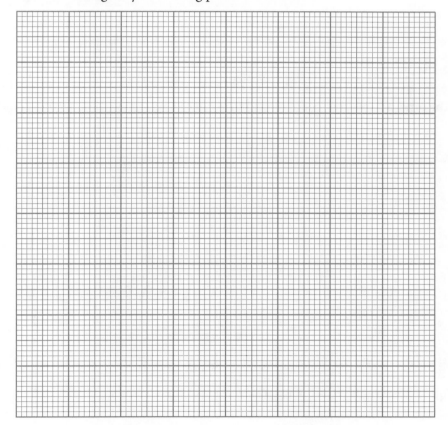

Conclusions

1 State the effect of physical activity on pulse rate.

....

2 Explain as clearly as you can, using your scientific knowledge, the reasons for the observations you made.

....

....

....

Evaluation

How could the measurement of your pulse rate be improved to make it easier for you, or more accurate?

....

....

....

GOING FURTHER

1 What are some of the other factors that could affect pulse rate? Suggest how these factors could affect pulse rate.

....

....

....

....

....

....

2 Plan an investigation to observe the effect of exercise on one of these factors.

....

....

....

....

....

Gas exchange in humans

9.1   Oxygen in exhaled air

We breathe in (inhale) to take in oxygen and breath out (exhale) to get rid of the carbon dioxide produced as a waste gas. When we breathe in we take in air, which contains about 78% nitrogen, 21% oxygen and 0.04% carbon dioxide. In this investigation we will compare the amount of oxygen in the inhaled air with the exhaled air.

Aim

To investigate the relative amount of oxygen in exhaled air.

Apparatus

- Eye protection
- Bowl of water
- Screw-top jar with two lids, one plain and the other with a candle holder
- Silicon tube
- Candle and wire holder
- Sterilising fluid
- Stopwatch
- Matches or lighter

SAFETY GUIDANCE

- Eye protection must be worn.
- The mouthpiece of the tube must be sterilised before reuse by another person. Refer to the CLEAPSS handbook and guidance for suitable steriliser fluids.
- Always take care with naked flames.

Method

Candle in ordinary air

1 Place the candle in the candle holder and light it.

2 Lower the candle into the jar and use the stopwatch to time how long the candle remains alight. Record the time below (see Figure 1c).

Candle in exhaled air

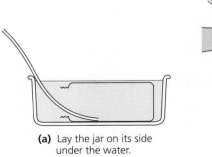

(a) Lay the jar on its side under the water.

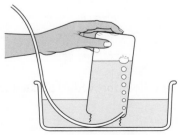

(b) Breathe out through the tube and trap the air in the jar.

(c) Lower the burning candle into the jar until the lid is resting on the rim.

Figure 1

3 Remove the lid from the jar and place the jar on its side in the bowl of water. Make sure it is completely full of water (see Figure 1a).

4 Put one end of the tube into the mouth of the jar, leaving the other end (mouth piece) out of the water (see Figure 1a).

5 Turn the jar upside down, ensuring the water and tubing remain inside.

6 Place your finger over the mouth piece tubing while you inhale, and then breathe out through the tube so the exhaled air collects in the jar and fills it (see Figure 1b). Take care not to inhale the water.

7 Place the screw-top lid onto the jar under the water, then sit it upright on the bench.

8 Light the candle in the wire holder.

9 Open the jar and quickly lower the candle in its holder into the jar (see Figure 1c).

10 Time how long the candle stays alight in the jar and record it below.

Observations

Record your measurements and observations below.

● Candle remained alight in ordinary air: ....

● Candle remained alight in exhaled air: ....

● Any other observations:

....

....

Conclusions

Explain your observations.

....

....

Evaluation

1 Did you expect what you observed? Explain your answer.

....

....

2 Outline how this experiment could be improved to make it more reliable or more accurate.

....

....

....

3 How do you think these results would compare if you had just finished carrying out some physical activity or exercise?

....

....

....

4 Plan an investigation to find out how exercise affects the amount of oxygen in exhaled air.

....

....

....

....

9.2   Carbon dioxide in exhaled air

Breathing is the process by which we take in oxygen and give out carbon dioxide. In this investigation we will test exhaled air and inhaled air with limewater.

Aim

To investigate the relative amount of carbon dioxide in exhaled air.

Apparatus

- Eye protection
- Delivery tubes and bungs prepared as indicated in Figure 1
- 2 large test tubes in a test-tube rack
- Limewater
- Silicon tube
- Sterilising fluid
- Stopwatch

Method

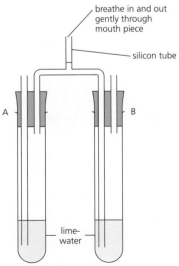

Figure 1

breathe in and out gently through mouth piece

silicon tube

A

B

lime-water

SAFETY GUIDANCE

- Eye protection must be worn.
- The delivery tubes are fragile. Take care when handling the apparatus.
- Limewater is an irritant to skin and eyes. Take care to avoid spillages. If it comes into contact with your eyes or skin, wash immediately.
- Avoid over-vigorous breathing.
- The mouthpieces must be sterilised before reuse by another person. Refer to the CLEAPSS handbook and guidance for suitable steriliser fluids.

1 Prepare two large test tubes, A and B, as shown in Figure 1.

2 Ensure that each tube contains a small amount of clear limewater.

3 Place the mouthpiece in your mouth.

4 **Gently** breathe in and out through the tubes for about 15 seconds. Notice which tube is bubbling when you breathe out and which one bubbles when you breathe in.

5 Record your observations.

Observations
What did you observe while breathing through the tube?

....

....

....

Conclusions
What can you conclude from your observations? Explain why there is a difference.

....

....

....

....

....

Evaluation
Explain why the tubes have been set up in this way.

....

....

....

....

....

GOING FURTHER

What factors might affect the rate at which the limewater changes appearance? How would you expect your results to change?

Outline how the experiment you have just done could be extended to investigate this.

....

....

....

....

....

....

....

....

....

Respiration

10.1 Investigating the effect of temperature on respiration in yeast

Yeast is a microorganism that can respire either aerobically or anaerobically. When oxygen is present, it respires aerobically; when oxygen is absent, it respires anaerobically. In both cases, it produces carbon dioxide.

Yeast has been used in baking for hundreds of years. The carbon dioxide gas helps to produce tiny holes in bread, making it soft and light. Respiration involves enzymes, and enzymes work best in certain conditions.

Aim

To investigate the effect of temperature on respiration in yeast.

Apparatus

- Eye protection
- Activated yeast solution (yeast in a warm sugar solution)
- Flour
- Source of warm and cold water
- Oil
- Tablespoon
- Teaspoon
- 4 × 250 cm³ beakers
- Petri dish
- Stirrer
- Thermometer
- 4 boiling tubes
- Permanent marker pen
- Ruler

KEY EQUATIONS

Aerobic respiration:

glucose + oxygen ⟶ carbon dioxide + water

Anaerobic respiration:

glucose ⟶ alcohol + carbon dioxide

KEY TERMS

aerobic
anaerobic
denature
enzyme
optimum temperature
respiration

SAFETY GUIDANCE

- Eye protection must be worn.
- This practical presents minimal risk. However, standard laboratory safety rules still apply at all times.

Method

1 Place 2 tablespoons of flour and 2 teaspoons of activated yeast solution in a Petri dish and mix to form a pliable dough.

2 Label 4 beakers with the appropriate temperature and set up 4 water-baths at a range of temperatures from 20–60 °C.

3 Swirl some oil inside each boiling tube so that the oil coats it (this stops the dough sticking to the glass and makes it easier to remove the dough after the experiment).

4 Use a stirrer to push a small piece of dough into the bottom of each of the boiling tubes, so that it is about a quarter full of dough.

5 Mark the height of the top of the dough on the boiling tubes using a permanent marker pen.

6 Check the temperature of each water-bath. Adjust the temperature by adding more warm or cold water to maintain the temperature.

7 Place a boiling tube into each of the water-baths. Leave them for at least 20 minutes. Keep checking the temperatures of the water-baths and adjust with hot water if required.

8 Draw a table to record the new height of the dough and calculate the rise.

Observations

Record the rise of the dough in a table here.

Conclusions

How did temperature affect the rate of respiration of the yeast? Explain your observations.

....

....

....

....

....

....

Evaluation

1 Suggest why sugar solution was added to the yeast.

....

2   Calculate the percentage change in height at each
    temperature.

KEY EQUATION

$$\text{percentage change} = \frac{\text{change in height}}{\text{initial height}} \times 100$$

3   Plot a graph of percentage change in height against temperature.

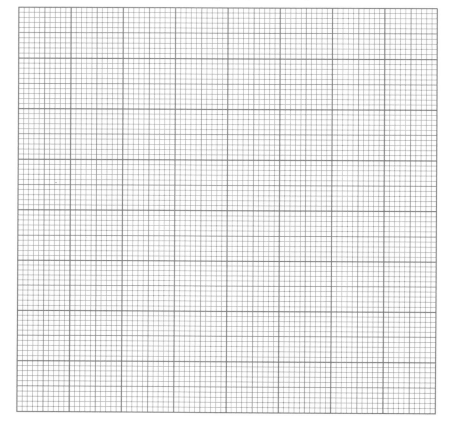

4   Which variables were controlled in this investigation?

....

5   Compare your results with those of other students in your class and suggest reasons for any differences.

....

....

....

6　Suggest why it is more useful to calculate the percentage change in height when comparing results with other students.

　...

　...

7　Outline how this experiment could be improved to make a more accurate estimate of the optimum temperature for the enzymes used in yeast respiration.

　...

　...

　...

8　The rate of respiration of yeast at different temperatures can be compared by measuring how much gas is produced in a specific amount of time. Outline a plan for an experiment using this method.

　...

　...

　...

　...

　...

9　Describe a method that could be used to show that the gas produced by respiring yeast is carbon dioxide.

　...

　...

　...

　...

Coordination and response

11.1   Investigating gravitropism in peas or beans

Gravitropism or geotropism is the influence of gravity on plant growth or movement. The response of seedlings to gravity can be investigated by putting some germinated peas or beans in a jar at different angles to test which way the root grows each time.

Aim

To illustrate gravitropism in the roots of peas or beans.

Apparatus

- 6 germinated peas or kidney beans with roots emerging
- Clear glass jar containing moist cotton wool
- Lid for the jar
- Different coloured marker pens

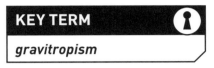

KEY TERM

gravitropism

SAFETY GUIDANCE ⚠️

- This practical presents minimal risk. However, standard laboratory safety rules still apply at all times.
- Wash your hands thoroughly after handling seedlings.

Method

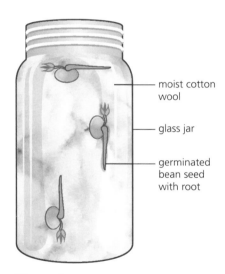

moist cotton wool

glass jar

germinated bean seed with root

Figure 1

1   Choose 4 seedlings with straight roots.

2   Carefully place the seedlings in the jar containing moist cotton wool, so that the seeds are pressed up against the glass. Place the seeds at equal intervals and position them so that each root points in a different direction: north, south, east and west (as shown in Figure 1).

3   Add about 1 cm of water to the bottom of the jar and put a lid on loosely.

4   Draw around the seedlings on the side of the jar with a permanent marker and/or take a photograph of each seedling.

5   Place the jar near a window for 2–4 days.

6   Draw around the seedlings with a different coloured marker.

7   Record your observations of each seedling by making drawings.

Observations

Record your observations below and describe the direction of growth of the seed roots.

Conclusions

Explain your observations.

....

....

....

Evaluation

1 Outline how this experiment could be improved or made more reliable.

....

....

....

2 Explain your observations in terms of the plant growth hormone auxin.

....

....

....

11.2 Investigating phototropism in shoots

Phototropism is the influence of light on plant growth or movement. The response of seedlings to light can be investigated by placing a box over some germinated pea or bean seeds and cutting holes in the box so that the direction that light gets into the box varies.

Aim

To illustrate phototropism in germinating seeds.

Apparatus

- 12 germinated mustard seeds or mung beans with shoots emerging
- 3 Petri dishes containing moist cotton wool
- Lids for the Petri dishes
- Different coloured marker pens
- Tape
- 3 cardboard boxes with lids:
    - Box 1 has holes cut out on all sides, including the lid
    - Box 2 has a hole cut out on one side to let light in from one side only
    - Box 3 has no holes cut out so no light can get in
- Access to a well-lit windowsill

KEY TERM

phototropism

SAFETY GUIDANCE

- This practical presents minimal risk. However, standard laboratory safety rules still apply at all times.
- Wash your hands thoroughly after handling seedlings.

Method

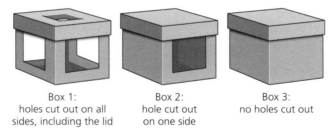

Box 1:
holes cut out on all
sides, including the lid

Box 2:
hole cut out
on one side

Box 3:
no holes cut out

Figure 1

1 Place four germinated seedlings on moist cotton wool in each Petri dish.

2 Cover each Petri dish with a lid. Draw around each seedling on the lid, and secure the lid with tape. Place one Petri dish in each box.

3 Place the boxes by a window so that:
    - Box 1 is exposed to light from all directions
    - Box 2 is exposed to light from one side, but the opening is facing away from the window
    - Box 3 is **not** exposed to light at all.

4 Mark each box so that you know the direction of light from the window.

5 Leave for 4–6 days and then record your observations by drawing and/or photographing the Petri dishes.

Observations

1   Record your observations for each box below.

2   How did the plants in Box 3 appear in comparison to the other plants?

....

....

....

Conclusions

Explain your observations using the concepts of photosynthesis and tropism.

....

....

....

....

....

Evaluation

1 In any investigation we need to keep some variables the same (constant) and need to change only one variable (the one we are measuring). Name the variable you changed in this investigation and list any variables you kept constant.

....

....

....

2 Which box acts as the control experiment?

....

3 Explain what a control is and why it is important to have a control for investigations.

....

....

4 Explain your observations in terms of the plant growth hormone auxin.

....

....

....

Reproduction

12.1 Insect-pollinated flowers

Reproductive parts of insect-pollinated flowers have evolved to attract insects to help pollinate the flowers. The reproductive parts can be observed by eye using a hand lens.

Aim

To identify the structures of an insect-pollinated flower.

Apparatus

- Insect-pollinated flower(s) from a local source, provided by your teacher
- Hand lens

Method

1 Observe a variety of insect-pollinated flowers using a hand lens.

2 Choose two different types of flowers and draw clear, large diagrams of your observations, labelling the petals, stamens, filaments, anthers, carpels, style, stigma, ovary and ovules.

3 Draw an annotated diagram to describe how the structure of a typical flower helps to attract insects. Make sure to label all parts of the flower listed above.

Observations

Record your observations here.

KEY TERMS 🔒

anther
filament
ovary
ovules
pollinate
stamen
stigma
style

SAFETY GUIDANCE ⚠️

- This practical presents minimal risk. However, standard laboratory safety rules still apply at all times.
- If you are allergic to pollen, inform your teacher before starting.
- Cover any cuts on your hands with waterproof dressings or wear gloves.
- Wash your hands thoroughly after handling plants.

Conclusions

List the features of the insect-pollinated flowers and describe how each feature helps to attract insects and aids pollination.

....

....

....

....

....

....

Evaluation

Research images of wind-pollinated flowers using books or the internet. Then draw a table to compare the structures of the insect-pollinated flowers you have observed with the structures of wind-pollinated flowers.

12.2 Germination: the need for water

Germination is affected by several environmental conditions. One condition that might be expected to affect germination is the availability of water.

Aim

To investigate the effect of water on the process of germination.

Apparatus

- 3 containers with lids
- Cotton wool
- Soaked pea seeds

Method

 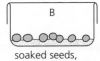

soaked seeds,
dry cotton wool

soaked seeds,
wet cotton wool

soaked seeds,
covered with water

Figure 1

1 Label the containers 'A', 'B' and 'C'.

2 Place dry cotton wool in the bottom of each container.

KEY TERM

control

germination

SAFETY GUIDANCE

- This practical presents minimal risk. However, standard laboratory safety rules still apply at all times.
- Wash your hands thoroughly after handling seeds.

3   Add an equal number of soaked seeds to each container.

4   Prepare the containers as follows:
    - A: cotton wool is left dry
    - B: cotton wool is moist
    - C: seeds are completely submerged in water.

5   Place lids onto the containers and leave at room temperature for a week.

6   Predict what you think will happen.

7   After a week, record your observations.

TIP

In any investigation, remember that all factors must be kept the same, except the one you are investigating.

Observations

1   Write a prediction of what you think you will observe.

....

....

....

2   Record your observations below.

....

....

....

Conclusions

How did water affect germination?

....

....

Evaluation

1   Name some factors that were kept the same (constant) in this investigation.

....

2   Outline how this experiment could be improved to make it more reliable.

....

....

....

12.3 Temperature and germination

Germination is affected by various environmental conditions. One condition that might be expected to affect germination is temperature.

Aim

To investigate the effect of temperature on the process of germination.

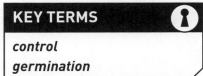

KEY TERMS

control

germination

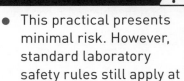

SAFETY GUIDANCE

- This practical presents minimal risk. However, standard laboratory safety rules still apply at all times.
- Wash your hands thoroughly after handling seeds.

Apparatus

- Soaked maize grains
- Blotting paper
- Water
- 3 opaque plastic bags
- Paper clips

Method

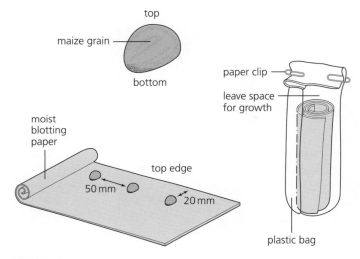

Figure 1

1 Cut three strips of blotting paper and moisten them.

2 Place an equal number of soaked maize seeds onto each strip.

3 Carefully roll up the strips of blotting paper and place each one into a plastic bag. Seal with paper clips (see Figure 1).

4 Place a bag in each of the following locations:
   - a refrigerator (at about 4 °C)
   - at room temperature (at about 20 °C)
   - on top of a radiator or in an incubator (at about 30 °C).

5 The breakdown of the food stored in the seed is dependent on enzymes. Using your knowledge of enzymes, make a prediction as to what you will observe.

6 Leave for a week and then examine the seeds. Record your observations.

Observations

1 Write your prediction here.

....

....

....

2 Draw a table and record your observations in the table.

Conclusions

1 How did temperature affect germination?

....

....

2 How can you explain this?

....

....

Evaluation

1 Suggest the reason for step 3 in the investigation.

....

....

2 Outline how this experiment could be improved to make it more reliable.

....

....

....

GOING FURTHER
• •

Plan an investigation into the growth of these plants by measuring increase in height over several weeks. What factors would you need to control that might affect growth over this time frame? Suggest how you would control each factor you have mentioned.

....

....

....

....

....

....

....

....

....

13 Variation and selection

13.1 Investigating variation between people in the class

Individuals in a population vary in many ways. Characteristics such as height, eye colour, skin tone and ability to roll your tongue are some observable variations in humans.

We can group these variations in terms of discontinuous variation and continuous variation. Discontinuous variation is usually caused by the genes you inherit. Continuous variation is caused by both genes and the environment.

KEY TERMS

bar chart
continuous variation
discontinuous variation
histogram

Aim

To collect data on physical characteristics of a group of people and decide whether each characteristic shows discontinuous variation or continuous variation.

Apparatus

● Tape measure

Method

1 Work in pairs. You will record the following characteristics in your classmates:
   – ability to roll the tongue (or not)
   – height
   – foot size (shoe size).

2 Measure your partner's height and foot (shoe) size, and observe whether or not they can roll their tongue.

3 Draw a table to record your results. Remember to include your own results.

4 Collect and add data from other groups in the class to your table.

5 Plot suitable graphs of the data collected in the space provided on page 104.

Observations

1 Draw a table in the space provided on the next page to record your observations.

SAFETY GUIDANCE

● This practical presents minimal risk. However, standard laboratory safety rules still apply at all times.

TIP

Discontinuous variation produces categoric data (in distinct categories). Each data set needs to be plotted as a bar chart. In a bar chart, the columns do not touch each other.

Continuous data can have any numerical value, so we use a continuous scale on the *x*-axis and plot the data in a histogram with grouped intervals. There are no gaps between the columns.

2 Consider how you will plot graphs of the data you have collected. Before you can plot the graphs, the data will need to be grouped and tallied. For example, you will need to count how many students in the class can roll their tongue and how many cannot, how many students have shoe size 3, size 4, etc. For variation in height you will need to group the data in continuous categories. It is best to use equal intervals, for example:
   – 151–155 cm
   – 156–160 cm
   – 161–165 cm, etc.

3 Draw another table with the data grouped and tallied.

4 Plot a graph of your tallied data for each characteristic.

Conclusions

1 What type of variation does the ability to roll your tongue show?

....

2 What type of variation do height and shoe size show?

....

Evaluation

1 Give a reason for the type of graphs you chose to use.

....

....

....

2 Describe the relationship, if any, between height and foot size.

....

3 Discuss why this relationship may have some exceptions.

....

....

....

4 Give four examples of other variable characteristics shown by members of your class. Classify these as
   continuous variation or discontinuous variation.
   a Examples of continuous variation: ....

   ....

   b Examples of discontinuous variation: ....

   ....

GOING FURTHER

Collect ten leaves from a tree. Record their length and plot a suitable graph of your data.

1 What type of variation did you observe in the leaves?

....

2 Discuss why the leaves on the same plant vary in shape, colour and size.

....

....

....

Biotechnology and genetic modification

14.1 The effect of pectinase on fruit pulp

Pectinase is an enzyme used to break down pectin, which is found in the walls of plant cells. Breakdown of pectin helps filtration, increases total juice yield and clarifies the fruit juice.

Note: This is not a required practical for the syllabus but teaches useful techniques.

KEY TERMS

cell wall
control
enzyme

Aim

To investigate the effect of pectinase on fruit pulp.

Apparatus

- Eye protection
- 2 stirrers
- Apple pulp
- 2 × 250 cm³ beakers
- 4 × 100 cm³ measuring cylinders
- 4 funnels
- Filter papers
- Stopwatch
- Balance
- Pectinase (follow manufacturer's guidelines for preparation, refer to safety instructions and refer to CLEAPSS guidance, see HC033 and GL116)

SAFETY GUIDANCE

- Eye protection must be worn.
- Wear a lab coat at all times. Handle all solutions with care.
- Avoid contact with skin, eyes or clothing. If solutions come into contact with your eyes or skin, wash immediately.
- Pectinase can cause irritation and allergies.
- Wipe up any spillages immediately and rinse the cloth thoroughly with water. Do not allow spillages to dry up.

Method

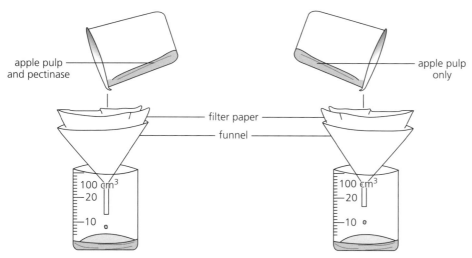

apple pulp and pectinase

apple pulp only

filter paper

funnel

100 cm³
20
10

100 cm³
20
10

Figure 1

1 Place 100 cm³ apple pulp in a 250 cm³ beaker.

2 Add 2 cm³ of pectinase enzyme (care needed – see safety note), stir the mixture and leave it for about 5 minutes.

3 Place the funnel in the top of a 100 cm³ measuring cylinder and line the funnel with a folded filter paper.

4 Transfer the pulp into the filter funnel and leave it in a warm place (e.g. near a radiator) for up to 24 hours.

5 Repeat steps 1–4 twice more, placing one measuring cylinder in the fridge and the other in a safe place outside the lab (e.g. windowsill).

6 Set up an identical control experiment without including pectinase.

Observations

1 Draw a suitable table to record the amount of fruit juice produced in each measuring cylinder.

2 Calculate the percentage difference for the amount of juice produced in each cylinder compared to the control experiment and record it in a new column in your table.

Conclusions

1 How did pectinase affect fruit juice production?

....

....

2 Compare the success of juice extraction at different temperatures.

....

....

3 Can you explain this? Relate your findings to your scientific knowledge.

....

....

....

....

Evaluation

1 Outline how this experiment could be improved or made more reliable.

....

....

....

2 What was the purpose of step 6 in the procedure?

....

....

GOING FURTHER

Pectinase can also be used to clarify fruit juice (make it more transparent). Enzymes are expensive and also difficult to separate from a food product. Juice manufacturers overcome this by placing the enzymes inside small alginate (inert gel) beads. The gel is partially permeable so the substrate and product can move in and out, even though the enzyme molecules cannot move. Once the juice has been produced, the gel beads can be retrieved and reused.

Plan an investigation showing how the method you used could be adapted to test whether gel beads containing pectinase could be used to clarify more than one batch of fruit juice.

....

....

....

....

14.2 Investigating the use of biological washing powders that contain enzymes

Protein and fats tend to be large insoluble molecules, which are difficult to remove from stained clothes. Many modern washing powders contain enzymes, and these are described as biological powders.

Aim

To investigate the use of biological washing powder compared to non-biological washing powder.

Apparatus

- Eye protection
- Egg
- Whisk, fork or stirrer
- $5 \times 250\,cm^3$ beakers
- Teaspoon
- 4 pieces of 10 cm × 10 cm cotton fabric
- $100\,cm^3$ measuring cylinder
- Cold water
- Boiling water
- Biological washing powder
- Non-biological washing powder
- Tongs

KEY TERMS

denature
enzyme
optimum temperature

SAFETY GUIDANCE

- Eye protection must be worn.
- Take care to avoid getting detergent on skin. If it comes into contact with your skin, rinse with water immediately.

Method

1   Break an egg into a beaker and whisk it with a fork, whisk or stirring rod until thoroughly mixed.

2   Smear egg evenly onto each of the 10 cm × 10 cm cotton fabric squares and leave to dry.

3   Label and set up four $250\,cm^3$ beakers as follows:
   - A: $100\,cm^3$ warm water with no washing powder.
   - B: $5\,cm^3$ (1 level teaspoon) of non-biological washing powder dissolved in $100\,cm^3$ warm water.
   - C: $5\,cm^3$ (1 level teaspoon) of biological washing powder dissolved in $100\,cm^3$ warm water.
   - D: $5\,cm^3$ (1 level teaspoon) of biological washing powder dissolved in $100\,cm^3$ water and boiled for 5 minutes, then left to cool until warm.

4   Place a piece of egg-stained cloth in each beaker and leave for 30 minutes.

5   Remove the pieces of cloth and compare the effectiveness of each washing process. Grade them from 0–3, where:

   0 = stain unaffected

   1 = stain slightly removed

   2 = stain almost removed

   3 = stain removed completely.

Observations

Record your observations below.

....

....

....

....

Conclusions

Explain your observations.

....

....

....

....

....

....

....

....

Evaluation

1   Explain why biological washing powders must be used at low temperatures.

....

....

2   Outline an investigation to make an accurate estimate of the optimum temperature for a biological washing powder.

....

....

....

....

3 Although biological washing powders are more expensive than non-biological washing powders, it is claimed that biological washing powders are more cost-effective to use, especially in washing machines. Explain why.

....

....

....

....

GOING FURTHER
• •

1 In this practical the results are based on a *qualitative* analysis, based on the appearance of the stains after washing. Suggest how you could modify the method to carry out the same investigation but collect *quantitative* results (results you can measure).

....

....

....

2 Biological washing powders are not recommended for washing silk and wool. Suggest why.

....

....

....

Past paper questions

Practical test past paper questions

1 Figure 1.1 is a photograph of a cross-section of a vascular bundle in a leaf.

Line **AB** shows the length of the vascular bundle.

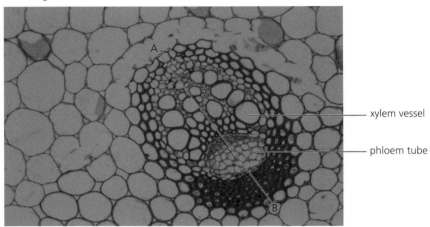

xylem vessel

phloem tube

Figure 1.1

a i) Make a large drawing to show the different regions of the vascular bundle shown in Figure 1.1.

Do **not** draw any individual cells.

Identify and label on your drawing the position of the xylem vessel as shown in Figure 1.1.   [5]

ii) Measure the length of line **AB** as shown on Figure 1.1. **Include the unit.**

Length of **AB** ....

Mark on your drawing a line in the same position as **AB**.

Measure the line you have drawn.

Length of line on drawing ....

$$\text{magnification} = \frac{\text{length of line on drawing}}{\text{length of } \mathbf{AB}}$$

Calculate the magnification of your drawing using the information above and your answers. Show your working.

magnification .... [3]

iii) State **one** way **visible** in Figure 1.1 in which the xylem vessel is different from the phloem tube.

....

....

.... [1]

b The walls of xylem vessels are supported by a chemical called lignin, which can be stained by a red dye. This makes the xylem vessel walls easily seen when using a microscope.

Use this information to plan how you could find the position of the vascular bundles in a stem.

....

....

....

....

....

....

.... [4]

[Total 13]

(Cambridge IGCSE Biology 0610 Paper 51 June 2016 Question 2)

2 You are going to investigate the effect of different concentrations of sucrose solution on the movement of water into and out of potato cells by osmosis.

Water enters cells if the solution outside the cells is less concentrated than the solution inside the cells.

Water exits cells if the solution outside the cells is more concentrated than the solution inside the cells.

You are provided with four potato sticks, which have been cut to the same length.

Read all the instructions but DO NOT CARRY THEM OUT until you have drawn a table for your results in the space provided in (a)(ii).

Step 1   Measure each of the potato sticks. Record the results in your table in **(a)(ii)**.

Step 2   Place one potato stick into each of the solutions in the large test-tubes labelled **A**, **B**, **C** and **D**. Immediately observe what happens to each of the potato sticks.

a i) Record your observations:

Potato stick in solution **A** ....

Potato stick in solution **B** ....

Potato stick in solution **C** ....

Potato stick in solution **D** .... [1]

Step 3   Leave the potato sticks for **30** minutes. While you are waiting continue with the other questions.

Step 4   Use a marker pen to divide the white tile into four sections and label them **A**, **B**, **C** and **D**.

Step 5   After 30 minutes pour the contents of the large test-tube **A** into the beaker labelled waste. Place the potato stick on to the white tile in the section labelled **A**.

Step 6   Measure the length of the potato stick and record the results in your table in **(a)(ii)**.

Step 7   Repeat steps 5 and 6 for large test-tube **B**.

Step 8   Repeat steps 5 and 6 for large test-tube **C**.

Step 9   Repeat steps 5 and 6 for large test-tube **D**.

ii) Prepare a table to record your results in the space provided.
Your table should show:
   – the length of the potato sticks at the start
   – the length of the potato sticks after 30 minutes
   – the change in length of the potato sticks.

[4]

iii) Pick up and examine each potato stick. State **two** physical differences, other than size, that you observe when comparing the four potato sticks.

1 ....

2 .... [2]

b i) Use all the information and your table of results to identify the solutions **A**, **B**, **C** and **D**. Write your answers in Table 2.1.

Table 2.1

relative concentration of sucrose solution	test-tube letter
least concentrated	
most concentrated	

[2]

ii) Explain how your results support your answer to part **(b)(i)**.

....

....

....

....

.... [3]

iii) Identify **one** source of error with the method and suggest an improvement.

error ....

....

improvement ....

....[2]

iv) State **one** of the controlled variables for this investigation.

....

....[1]

c Another investigation was carried out into the effect of different concentrations of sucrose solution on potato sticks.

In this investigation students decided to measure the change in mass rather than the change in length.

The students followed a similar method to the one in your investigation but they left the potato sticks to soak for three hours instead of 30 minutes.

i) Suggest why the students left the potato sticks in the solutions for three hours instead of 30 minutes.

....

....

....[1]

ii) The students dried the potato sticks on paper towels before measuring the mass of each potato stick.
Suggest why this step was **not** important in your investigation, where length was measured.

....

....

.... [1]

Table 2.2 shows their results.

Table 2.2

concentration of sucrose solution/ g per dm³	percentage change in mass
0	29.5
70	12.0
140	−3.0
210	−15.0
280	−26.0
350	−29.5

iii) Using Table 2.2, plot a graph on the grid to show the effect of the concentration of sucrose solution on the percentage change in mass.
The *y*-axis has been started for you.

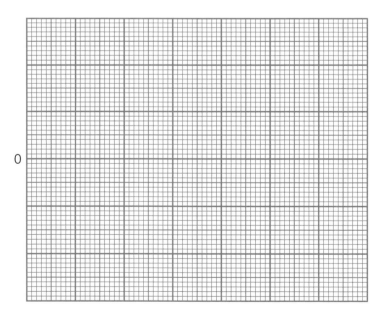

[4]

iv) Use your graph to find the concentration of sucrose solution that would cause **no change** in mass of the potato stick.
Mark this concentration on your graph with a **+** and record the concentration.
Include the unit.

.... [2]

v) Students tested other potatoes and found different values for the concentration of sucrose solution that would cause no change in mass.

Suggest **one** reason for this.

....

....

.... [1]

[Total: 24]

(Cambridge IGCSE Biology 0610 Paper 52 March 2017 Question 1)

3 Young mammals feed on milk. Milk contains protein.

Some mammals produce an enzyme called rennin. Rennin changes the protein in milk so that it can be digested by another enzyme.

The action of rennin causes small lumps or clots to form in the milk.

You are going to investigate the effect of pH on the activity of the enzyme rennin.

Read all the instructions but DO NOT CARRY THEM OUT until you have drawn a table for your results in the space provided in (a).

Use the gloves and eye protection provided while carrying out the practical work.

Step 1	Label three test-tubes **P**, **Q** and **R**.
Step 2	Use a syringe to add 5 cm^3 of milk into each of test-tubes **P**, **Q** and **R**.
Step 3	Add two drops of acid to test-tube **P**.
Step 4	Add two drops of distilled water to test-tube **Q**.
Step 5	Add two drops of alkali to test-tube **R**.
Step 6	Label another three test-tubes **P1**, **Q1** and **R1**.
Step 7	Use a clean syringe to add 1 cm^3 of 0.1% rennin solution into each of test-tubes **P1**, **Q1** and **R1**.
Step 8	Raise your hand when you are ready for water to be added to the beaker labelled **water-bath**.
Step 9	Place all six test-tubes into the filled water-bath and leave them for three minutes.
Step 10	Pour the contents of test-tube **P1** into test-tube **P**.
	Pour the contents of test-tube **Q1** into test-tube **Q**.
	Pour the contents of test-tube **R1** into test-tube **R**.
Step 11	Leave test-tubes **P**, **Q** and **R** in the water-bath.
	Immediately start the stop-clock.
	The empty test-tubes, **P1**, **Q1** and **R1** can be placed in the test-tube rack.
Step 12	After one minute, take test-tube **P** out of the water-bath.
	Tip and rotate test-tube **P** as shown in Figure 3.1.
	Observe the milk, and decide which stage of clotting (**no clotting**, **some clotting** or **all clotted**) it has reached.
	Record your result in your table.

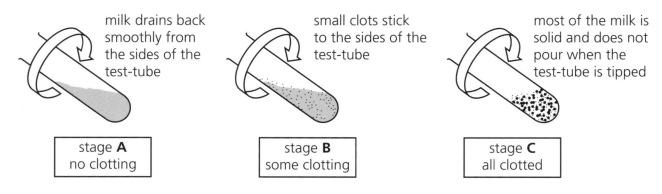

Figure 3.1

Step 13   Put test-tube **P** back into the water-bath.

Step 14   Repeat steps 12 and 13 for test-tubes **Q** and **R**.

Step 15   Repeat steps 12, 13 and 14 every minute for five minutes.

a Prepare a table in which to record your results.

[4]

b State a conclusion for your results.

....

....

....

....

.... [2]

c **i)** Suggest why, in step 9, all the test-tubes were placed into the water-bath for three minutes before mixing the contents together in Step 10.

....

.... [1]

ii) State **two** variables which have been kept constant in this investigation.

1 ....

2 .... [2]

d Identify **four** sources of error in this investigation.

1 ....

....

2 ....

....

3 ....

....

4 ....

.... [4]

e Identify **one** hazard associated with this procedure that required you to wear eye protection.

....

....

.... [1]

f Clotting separates milk into a solid part and a liquid part.
Describe how you could find out if there was any protein remaining in the liquid part.

....

....

....

.... [2]

g After rennin has changed the protein in milk into a white solid, protease enzymes can be used to digest the protein. The digested protein forms a colourless liquid.

A hypothesis stated:

The optimum temperature for protease enzymes to digest changed milk protein is 37 °C.

Describe a method that could be used to test this hypothesis.

Do **not** carry out this investigation.

....

....

....

....

....

....

.... [6]

[Total: 22]

(Cambridge IGCSE Biology 0610 Paper 51 June 2018 Question 1)

Alternative to Practical past paper questions

1 Some students test the composition of three liquid food supplements.

 a i) State the chemical test the students would use to show that protein is present in a liquid sample of a food supplement.

 .... [1]

The students carried out this test for protein on liquid samples of food supplements **P**, **Q** and **R**.

Food supplements **P** and **R** contained protein.

 ii) Complete Table 1.1 to show the results from the students' tests for protein. [2]

Table 1.1

food supplement	colour at start	colour at end
P		
Q		
R		

The students carried out a test for vitamin C on liquid samples of food supplements **P**, **Q** and **R**.

When iodine solution is mixed with starch, a blue-black colour is observed. Vitamin C stops this blue-black colour from forming.

Step 1   The students labelled a test-tube **P** and added $3\,cm^3$ of food supplement **P** to the test-tube.

Step 2   They added $1\,cm^3$ of starch solution to test-tube **P**.

Step 3   The students added iodine solution to the test-tube, one drop at a time. They counted the drops as they added them. They shook the test-tube gently after adding each drop and stopped adding drops when a blue-black colour remained.

         A blue-black colour remained in **P** after **12** drops of iodine solution had been added.

Step 4   They repeated steps **1** to **3** with food supplements **Q** and **R**.

         A blue-black colour remained in **Q** after **1** drop of iodine solution had been added.

         A blue-black colour remained in **R** after **5** drops of iodine solution had been added.

Table 1.2 shows how the number of drops of iodine solution added relates to the vitamin C content of the food supplement.

Table 1.2

number of drops of iodine solution added	vitamin C content
1	none
2–3	low
4 or more	high

b Use the results of the students' experiments and the information in Table 1.2 to complete Table 1.3.

Table 1.3

food supplement	number of drops of iodine solution added	vitamin C content
P		
Q		
R		

[2]

The students carried out a test for reducing sugar on liquid samples of food supplements **P**, **Q** and **R**.

c i) Name the solution used for the reducing sugar test.

.... [1]

ii) Give **one** safety precaution that should be used when carrying out this test.

.... [1]

A positive result for the test for reducing sugar is the appearance of a brick-red colour.

The quicker the brick-red colour appears, the higher the concentration of reducing sugar.

Step 5   The students labelled a test-tube **P2** and added a sample of food supplement **P** to the test-tube.

Step 6   They added $2\,cm^3$ of the test solution to test-tube **P2**.

Step 7   The students repeated steps **5** and **6** with food supplements **Q** and **R**.

Step 8   They placed test-tubes **P2**, **Q2** and **R2** into hot water, and started a timer.

Step 9   The students observed the test-tubes carefully and noted the time when the brick-red colour appeared in each test-tube.

If there was no colour change after 180 seconds (3 minutes), the students recorded 'more than 180' as the result for that test-tube.

A brick-red colour appeared in test-tube **R2** after 25 seconds and in test-tube **P2** after 1 minute and 15 seconds.

No brick-red colour appeared in test-tube **Q2**.

d Complete Table 1.4 to show the students' results for the reducing sugar test.

Table 1.4

test-tube	time for brick-red colour to appear/s

[2]

e There is a source of error in step **5** of the method for the reducing sugar test.

i) Identify this source of error.

....

....

.... [1]

ii) Suggest apparatus that could be used to minimise this source of error. [1]

....

f State **one** other source of error in the method used for the reducing sugar test.

Suggest how to improve the method to minimise this source of error.

error ....

....

improvement ....

....

.... [2]

g Table 1.5 shows the protein content of five foods.

Table 1.5

food	protein content of food/g per 100 g
maize	3.2
rice	7.1
potato	2.0
yam	1.5
sorghum	11.3

i) Plot a graph of the data shown in Table 1.5.

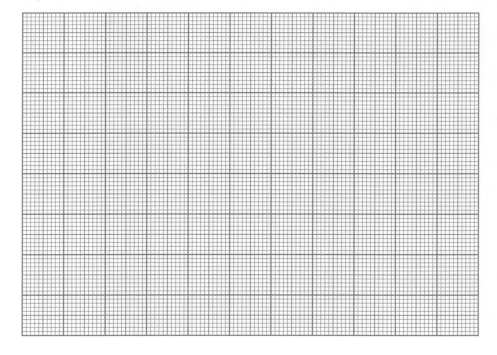

[4]

ii) It is recommended that a six-year-old child eats 20 g of protein per day.

Calculate the mass of sorghum a six-year-old child must eat each day to obtain 20 g of protein.

Show your working.

Give your answer to the nearest whole number.

....g

[2]

[Total: 19]

(Cambridge IGCSE Biology 0610 Paper 62 June 2016 Question 1)

2 Figure 2.1 is a photomicrograph of some blood cells.

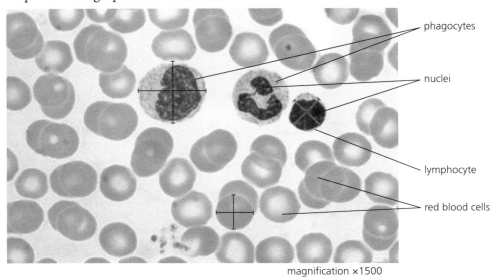

magnification ×1500

Figure 2.1

a i) State **two** visible differences between the red blood cells and the white blood cells (phagocytes and lymphocytes) in Figure 2.1.

1 ....

....

2 ....

.... [2]

ii) Make a large drawing of the two cells labelled **phagocytes** in Figure 2.1. [4]

b i) Measure the diameters of the three marked blood cells, along both the lines drawn on each of the cells, in Figure 2.1. Record these measurements in Table 2.1.

Add the missing units to Table 2.1.

Calculate the average diameter for each type of blood cell and write your results in Table 2.1.

Table 2.1

type of blood cell	diameter 1 / ……..	diameter 2 / ……..	average diameter / ……..
red blood cell			
lymphocyte			
phagocyte			

[3]

ii) Calculate the actual average diameter of the red blood cell using your answer in **2(b)(i)** and the following equation.

$$\text{magnification} = \frac{\text{average diameter of the red blood cell in Figure 2.1}}{\text{actual average diameter of the red blood cell}}$$

Give your answer in micrometres (μm) to the nearest whole number. 1 mm = 1000 μm

Show your working.

.... μm

[3]

[Total: 12]

(Cambridge IGCSE Biology 0610 Paper 61 November 2017 Question 2)

3 A student measured the distance moved by different concentrations of citric acid solution through agar jelly.

The agar contained Universal Indicator which changed colour in the presence of acid. The agar mixed with Universal Indicator was green at the beginning of the investigation.

Step 1   Three test-tubes were labelled **A**, **B** and **C**. Three different concentrations of citric acid solution were made.

Table 3.1 shows the volumes of 5% citric acid solution and distilled water that were used to make each solution.

Table 3.1

	solution		
	A	**B**	**C**
volume of 5% citric acid solution/cm³	1.0	2.0	10.0
volume of distilled water/cm³	9.0	8.0	0.0
percentage concentration of citric acid solution	0.5	1.0	5.0

Step 2   The base of a Petri dish containing agar and Universal Indicator was labelled **A**, **B** and **C**.

Three holes were cut into the agar. This is shown in Figure 3.1.

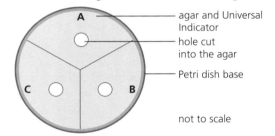

Figure 3.1

Step 3   The student was provided with one dropping pipette. Three drops of solution **A** were placed into the hole in section **A** of the Petri dish.

Step 4   Three drops of solution **B** were placed into the hole in section **B** of the Petri dish.

Step 5   Three drops of solution **C** were placed into the hole in section **C** of the Petri dish.

Step 6   A stop-clock was started.

Step 7   After 30 minutes the student observed the colour change in the agar around the hole in each section of the Petri dish. The colour change was caused by the diffusion of the citric acid solution through the agar.

Step 8   A ruler was used to measure the distance travelled by each concentration of citric acid solution through the agar.

Figure 3.2 shows the appearance of the Petri dish after 30 minutes.

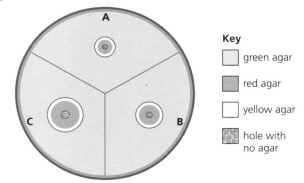

Figure 3.2

a Use a ruler to measure the distance travelled by each concentration of citric acid solution after 30 minutes in Figure 3.2.

Record these results in your table in **3(a)(i)**.

i) Prepare a table to record the results.

You should include:
- the concentration of the citric acid solutions
- the distance travelled by the citric acid solutions.

[3]

ii) Describe how you decided where to measure the distance travelled by the citric acid solutions.

....

....

.... [1]

iii) State a conclusion for these results.

....

....

.... [1]

iv) The citric acid moves through the agar by diffusion. The diffusion coefficient is used to show the effect of concentration on diffusion.

The formula to calculate the diffusion coefficient is:

$$\text{diffusion coefficient} = \frac{(\text{distance travelled})^2}{\text{time}}$$

Calculate the diffusion coefficient for a 10% solution of citric acid that travelled 14 mm in 30 minutes.

Give your answer to two significant figures.

.... mm² per minute [2]

v) Universal Indicator is used to estimate the pH value of substances.
Estimate the pH value for the green agar and the red agar.

green agar pH ....

red agar pH .... [2]

b i) State **two** variables that have been kept constant in this investigation.

1 ....

2 .... [2]

ii) Identify **one** potential source of error in this investigation and suggest how the error could affect
the results.

error ....

....

effect on results ....

.... [2]

c Describe how you could adapt this method to find the effect of temperature on the rate of diffusion.
Agar melts at 70 °C.

....

....

....

....

....

....

....

.... [6]

[Total: 19]

(Cambridge IGCSE Biology 0610 Paper 61 June 2019 Question 1)